New concepts and approaches to land management in the tropics with emphasis on steeplands

75

by
Francis Shaxson
FAO Consultant

Soil Resources, Management and Conservation Service
Land and Water Development Division

Food
and
Agriculture
Organization
of
the
United
Nations

Rome, 1999

M-57
ISBN 92-5-104319-1

Foreword

Small farmers living on steeplands in the tropics comprise a large sector of the world population. They operate in subsistence agricultural systems using traditional technologies designed to achieve sustained crop production at very low yield and income levels. Thus, most steeplands have relatively stable, sustained production systems with minimum land degradation. However, there is a growing awareness that sustainability of subsistence agriculture on steeplands is steadily deteriorating as a result of rapid growing population and overexploitation of the land resource base. Given the various limitations affecting steeplands agriculture, few soil management and conservation options remain. This bulletin attempts to explore potential avenues for developing strategies to maintain sustainability of steeplands agriculture. To maintain sustainability, however, significant improvements in crop production on steeplands must come from improved soil management practices focusing on soil and water conservation, improved crop management, and use of cropping systems designed to reduce soil erosion and maintain soil productivity.

Acknowledgements

The author is grateful to José Benites and David Sanders, AGLS, for the advice on preparing this publication and their direction, support and technical guidance. Recognition is given to a previous document entitled "Soil Management and Conservation for Tropical Steeplands" by Luis A. Manrique, University of Hawaii, which provided valuable inputs for the present text.

Special thanks are due to Richard Barber for his important contribution to this document and for the final editing. Also valuable were the inputs and proof reading provided by Robert Brinkman.

Finally, the author thanks the many enthusiastic farmers and technicians practising better land husbandry worldwide, whose knowledge has helped to lay the foundation for this publication and, last but not least, to Lynette Chalk for her efficient preparation of the text and formatting of this document and Chrissi Redfern for the final copy editing.

Contents

List of tables

List of boxes

List of figures

Chapter 1

Introduction

CONCERNS ABOUT STEEPLANDS

Most of the land degradation problems found in steeplands also occur in gently sloping areas, but the difficulties and limitations of managing lands productively, safely and cheaply become more severe as slope gradient increases. In particular, the steeper the slope the more rapid are the land degrading processes of soil erosion, runoff and mass movement due to the effect of the earth's gravitational field. This implies that any mistakes in land management will have more rapid and potentially more damaging consequences on steep slopes, than would be the case on gentler slopes. This is one of the prime reasons why steeplands are generally considered to be 'ecologically fragile'.

A corollary of the greater susceptibility of steeplands to land degradation compared to more gently sloping areas, is the lower capacity of the former to absorb rainwater compared to the latter. Thus crop-water availability is generally more critical in steeplands than in nearby flat lands within the same climatic zone.

Many steepland areas in the tropics are already inhabited, and present trends indicate that more of the steeplands will become inhabited in the future, and that population densities will increase. The majority of inhabitants are resource-poor, small-scale, farming families whose livelihoods may be seriously jeopardized by land degradation and associated loss of productivity.

In such situations, there are few, if any, opportunities for governments to move the people to more gently sloping areas where the problems of land management are less acute.

Many aspects of disquiet exist about steeplands:

- Legislation has generally been unsuccessful in preventing people from occupying steeplands.

- Even though settlement of steeplands is increasing, in some regions fields are being abandoned from annual cropping.

- Common recommendations about soil conservation have generally not been readily adopted or maintained by farmers. When soil conservation measures have been applied, they do not appear to have adequately solved the problems of soil erosion on their own, nor to have given the anticipated yield increases – there is evidence that their use may even reduce agricultural output (Box 1). This may be due to the exposure of less productive subsoils during terrace formation and/or to taking strips of land out of the preferred form of production to create the conservation measures. The steeper the land, the closer the conservation measures are spaced, and hence the greater will be the effect.

BOX 1:   **EFFECTS OF SOIL CONSERVATION WORKS ON OUTPUT/HA
FOR CROP YIELDS IN ETHIOPIA**

In a soil conservation research programme in Ethiopia, results from four locations showed that
the imposition of soil conservation measures, of one to four years of age at the date of
measurement, resulted in decreased yields.

Location	Crop	Control plot	Grass strip	Level bund	Level *fanya-juu*	Graded bund	Graded *fanya-juu*
				Yield (kg/ha)			
Wello	Maize	1860	1380	1660	510	930	1420
Omo	Sorghum	780	680	560	580	680	780
Shewa	Peas	170	140	180	170	220	130
Shewa	Beans	1000	520	820	640	800	480

Source: Herweg and Grunder (1991), quoted by Herweg (1993:405).

- Fifty years of soil conservation research have still not managed to establish clear quantitative relationships between soil losses and crop yields.

- Bush fallows, which served to maintain low levels of productivity in areas with low population densities, are becoming so short as to become virtually worthless in this role.

- Even the rectification of faults in the design of soil conservation projects seems to be inadequate to ensure the adoption and maintenance of soil conservation measures; furthermore, most detailed technical manuals on soil conservation provide insufficient instructions on how to ensure that recommendations are adopted.

- Politicians and other decision-makers at the head of large institutions and government agencies appear relatively unconcerned about the low success rate in halting erosion and raising yields on steeplands and elsewhere.

The situation in steeplands in the tropics is sufficiently serious to warrant a re-assessment in accordance with the spirit of Chapter 10 of Agenda 21, resulting from the UN Conference on the Environment in Rio de Janeiro.

OBJECTIVES OF THE BULLETIN

This *Bulletin* has several related objectives:

- to present alternative ideas on the management of steeplands, and in particular on the improvement of soil quality which will enhance rainwater infiltration and accessibility of plant nutrients and lead to beneficial changes in the type of management practices implemented on steeplands;

- to modify some common assumptions, preconceptions and perceptions about farmers and their problems of declining land productivity;

- to encourage extensionists and research workers to adopt participatory approaches when working with farming families;

- to dispel some of the doubts concerning the possibilities of improving land management and achieving greater and more sustained productivity from steeplands;

- to contribute to more effective agricultural development procedures that will result in greater food security and better living standards in tropical steepland regions.

This *Bulletin* emphasizes how to improve and maintain soil quality, in participation with resource-poor, small-scale farmers, rather than how to conserve soil and water *per se*, on which many textbooks and manuals already exist.

It is postulated that soil and water conservation is more effectively, sustainably and acceptably achieved via improvements in the husbandry and management of crops, animals and land than by relying on the addition of conservation-specific practices to the production systems.

So as to be of the widest applicability, the *Bulletin* emphasizes principles which may be applied in most if not all situations. The translation of the principles into field practice will require practical actions which are well adjusted to the site-specific and farmer-specific conditions where they are to be applied.

It is believed that if stability and sustainability of production can be achieved on steeplands, then it should be possible to achieve the same more easily on lands of lesser slope.

ORGANIZATION OF THE DOCUMENT

The first chapter deals with present day concerns about the problems associated with steeplands in the tropics, and the objectives of the document. The definition, occurrence and nature of steeplands and their associated climate, soils and vegetation are briefly dealt with in Chapter 2. This is followed in Chapter 3 by a consideration of the socio-economic, political and environmental problems, with emphasis on soil degradation processes that arise when steeplands are settled and arable agriculture is introduced. Past concepts and approaches to solving steepland problems are considered in Chapter 4, and the changes in concepts and approaches required for improved steepland management, which is the central theme of this document, are presented in Chapter 5. Practical applications of these new concepts and approaches to the planning, management and monitoring of steeplands are presented in Chapter 6, and their implications in Chapter 7.

READERSHIP

This document is directed to those who are strategic thinkers, decision-makers and planners of agricultural development programmes in the tropics, as well as practitioners, such as extensionists, project managers, researchers and farmer-leaders who work with resource-poor, small-scale farmers.

It is assumed that readers will have some technical knowledge in their particular discipline and field of interest and, in addition, will be interested and disposed to work in inter-disciplinary ways with other specialists in the process of helping farmers solve their own problems.

Chapter 2

The nature of steeplands in the tropics

DEFINITION OF STEEPLANDS

'Steeplands' is a relative term, and refers to those lands that exceed a specified slope limit within a continuum of slope gradients. The limiting slope gradient, beyond which arable production is considered 'unsuitable' in land classification systems, varies from region to region. In Central Africa the limit has generally been set at 12%, in Israel the limit is 15%, whereas in the Philippines, 25% is considered the maximum slope unless there is slope amelioration (Hudson 1981:173). In Taiwan a 30% slope is tolerated for arable use, but "very intensive conservation practices [are] necessary" (Liao and Wu 1987:12, 13).

Thus there is no single, uniform slope limit for arable production, and the accepted limits for annual crops appear to be greater in areas with limited extensions of gently sloping land, and where many people are already farming steep slopes.

For the purposes of this document, 'steepland' is defined as land with an average slope of more than 12%, which is approximately equal to 7° (Annex 1 presents the relationships between slope percentages and degrees). At 12% slope it is dangerous to use normal four-wheeled tractors along the contour, because of the risks of overturning, and so this slope limit determines whether or not mechanized tillage and weeding can be undertaken with conventional machinery.

DELINEATION OF STEEPLANDS AND SLOPE CLASSES

Topographic maps may be used to delineate the approximate distribution of steeplands in an area, as well as the distribution of different slope classes relevant to land management. However, field checking to verify the acceptability of the precision of the estimated slope classes is always necessary. A procedure to delineate steeplands and different slope classes on topographic maps is presented in Annex 2. For mapping at the farm level, map scales must be meaningful in relation to farm size: generally 1:25 000, 1:10 000 or more detailed scales.

OCCURRENCE OF STEEPLANDS

In both wet and dry tropical zones, lands of more than 30% slope constitute 17% of the total land area (Buol and Sanchez 1988). Therefore steeplands with slopes greater than 12% will occupy an even larger proportion, perhaps more than 20%.

Of the 1 700 million hectares in the Asian region, about 14% have slopes steeper than 30%, and a further 39% have slopes of 8-30%. About 65% of the region's rural population of 1 600 million live in steeplands (Doolette and Magrath 1990:1). In contrast, in Brazil, there is a very small proportion of land with slopes greater than 12%.

Steeplands occur in many different land types, and become more apparent on maps of larger scales. For example, low, steep-sided hills which would not be represented on a map of 1:250 000 scale, would be clearly apparent on a contoured map of 1:25 000 scale.

GEOLOGICAL PROCESSES AND THE FORMATION OF STEEPLANDS

The form of the land as seen today is a result of continuing slow geological processes of uplift and formation of new land masses, their wearing down by wind and water erosion, the removal of eroded materials by wind, water, gravity, and in a few places, by glaciers.

Steeplands may form by the deposition of ash around volcanic vents, giving characteristic conical shapes, such as Mt. Fuji in Japan and Cotopaxi in Ecuador. They may result from movements associated with geological faults, such as the Rift Valley in eastern Africa, the Zagros fault in western Iran, and the Drakensberg escarpment in Lesotho. Mountain-building actions of tectonic plate movements, such as those responsible for thrusting up the Himalayan ranges, the Andes, or the South Island of New Zealand, result in steep slopes not only by uplift and folding, but also by mass movements, e.g. landslides.

Where the geomorphology of an area is not affected by volcanic eruptions or tectonic movements, processes of rock weathering, geological erosion and sediment transport may slowly expose harder rocks that give rise to steep slopes as a result of their greater resistance to weathering compared to surrounding rocks. For example, where sedimentary rocks or very fluid volcanic lavas have been laid down in layered horizons of contrasting hardness, and which are now either level-bedded or tilted, short sections of steep slopes may occur below the hard rock bands where the intervening softer materials have been worn away.

Mass movements such as landslides and mud flows are specific to steeplands. They may be triggered by seismic activity or by exceptionally prolonged and heavy rainstorms, and may be caused by the undercutting of riverbanks, or by excavations into hillsides for quarrying or road construction. Mass movements are also common in areas where there have been geologically-recent changes in the hydrological base level due to up-thrusting of the land mass.

A study in the Middle Hills of Nepal showed that debris slides were generally unrelated to human activity, whereas small-scale slumps were often triggered by surface saturation in terraced, irrigated land (Gardner and Jenkins 1995:8). Landslides, occurring relatively infrequently, whether on a large or small proportion of a steep landscape, often contribute most of the sediment found in upland river systems.

Although it is rare to see a landslide, dust storm or soil erosion in action, these geological processes are as active today as they were in the past, and will continue to drive landscape formation in the future. These processes will be better understood by venturing out and witnessing them directly (Shaxson 1971:193).

If a land surface is abraded by wind or water at the rate of 1 mm per year, the surface would be lowered by 1 000 m (1 km) in 1 million years, a rate that would be virtually imperceptible to the human eye, but which is relatively rapid on the geological time-scale.

In spatial terms, every topographic change, even on a small field, represents a change in the shape of the landscape; major changes in a landscape are the summation of all these small-scale topographical changes in small areas over long periods of time.

ASPECT

Insolation of the land surface is strongest on slopes which directly face the sun, i.e. south-facing slopes in the northern hemisphere, and north-facing slopes in the southern hemisphere. Between the Tropics of Cancer and Capricorn (23°30' N and 23°30' S of the equator), the sun is overhead twice a year, and all slopes receive some insolation, but steep slopes facing away from the sun at a given time will only receive oblique insolation. In mountainous regions, flanks of mountains may cast shade into valleys, limiting the hours of direct sunlight and lowering mean daily temperature in the shaded area. Aspect may also affect the exposure to prevailing winds and rainfall.

Thus within a region of complex slopes, there may be not only sun-shadow but also wind-shadow and rain-shadow effects (Hallsworth 1987:20). When these effects are combined with differences in soil type, a multiplicity of different ecological niches can develop within a landscape. This is an important feature of steeplands; the resultant heterogeneity produces a range of environments which may require contrasting systems of land management, but also create possibilities for a high degree of agricultural diversification.

CLIMATE

Climate - elevation interactions: For every 100 m increase in altitude there is a cooling of approximately 1°C, which is known as the "adiabatic lapse rate". At the equator the snow line is at its greatest altitude (ca. 4 200 m), whereas at the Poles it is at sea level. Thus the further from the equator, the lower will be the mean temperature for a given altitude above sea level.

In mountainous areas with large differences in elevation, moving air masses cool as they are forced upwards, and often reach the point of condensation. Thus, much more precipitation generally occurs on the upper parts of windward slopes of high hills and mountains than on the leeward slopes. Conversely, as the air masses pass the crest and begin to move down slope, they warm, and having already precipitated much of their rain, pass down slope releasing only small quantities of additional rain (Schwab *et al.* 1966:30). For these reasons, windward slopes in highland locations are important source-areas for streams (Pereira 1973:25), whereas leeward slopes experience much lower precipitation due to the 'rain-shadow' effect (Plate 1). This can greatly influence land use. For example, in southern Malawi, the main area of the tea estates is concentrated on the southern windward flank of Mulanje mountain because of the higher rainfall and the high moisture requirement of tea. In some countries of the Pacific region, rainfall on windward slopes may exceed 10 000 mm per year, while leeward slopes may receive no more than 50 mm per year (Willatt 1993:196).

At high altitudes, lower temperatures slow down the rates of plant material decomposition, mineralization of soil organic matter, weathering, soil-formation processes, evaporation from moist soil surfaces, and evapotranspiration. Consequently, plant growth rates are reduced, and the length of growing period is less than it would be at lower altitudes (FAO 1991:27, 28). At high elevations in the tropics, the limitations of cold soil conditions are barely influenced by seasons, and soil temperatures vary little throughout the year (Buol and Eswaran 1993:18).

In hilly and mountainous regions, local topography may significantly influence rainfall causing substantial variations over short distances of only a kilometre or so. Farmers may have detailed knowledge of the wetter and drier spots in their localities, which are the result of the direction of prevailing rain-bearing winds, rain-shadow effects and local wind-deflection by significant hill and mountain masses.

PLATE 1
The warmer and wetter windward slope of the hill is favoured for farming and for trees because of the greater rainfall. Nepal
[Natural Resources Institute]

Rainfall reliability: In relation to annual crops, rainfall reliability at different times of the year is very important because it influences cropping risks, grain yield and dry matter production. For these reasons it is often necessary to evaluate the reliability of rainfall at a given location using ten-day periods, but there should be at least ten years of daily rainfall records available. Annex 3 shows how to calculate the reliability or probability of rainfall over each dekad (ten-day unit) of the year.

Rainfall intensity and energy: Unlike rainfall in temperate regions, much of the rain in the Inter-Tropical Convergence Zone (which girdles the earth between about 30°N and 30°S) falls at intensities above 25 mm/hr, the general threshold intensity above which raindrops are erosive (Hudson 1995: Chapter 3). Rainfall intensity is not closely related to total rainfall; very high rainstorm intensities occur in semi-arid areas, but total rainfall energy may be greater in high rainfall areas because of a greater number of erosive rainfall events that are individually of less erosive power than the few highly erosive events occurring in lower rainfall areas. In highland areas much of the annual precipitation may occur as cloud and mist, with very low and non-erosive intensities. Annex 4 presents the procedure for estimating rainfall energy and erosivity when autographic rain gauge data are available. See also Hudson 1995: Chapters 3 and 4.

Length of growing period: The length of growing period (LGP) is defined as the duration (in days) when both water and temperature are adequate for crop growth. A moisture supply of half, or more, of the potential evapotranspiration is considered sufficient for plant growth when mean daily temperatures exceed 5°C (Higgins *et al.* 1982:12 and maps). In limited areas of the tropics the LGP may be as long as 365 days per year, but in most areas the LGP is of much shorter duration. In steeplands, the limitations of dry periods, low soil-moisture holding capacities due to shallow and stony soils, and lower temperatures at higher altitudes, are likely to be the main factors limiting the length of growing period.

SOILS

Parent materials: The types of soil parent material in steeplands are as varied as elsewhere, but because of geological folding and faulting in hilly and mountainous areas, rapid changes in the parent material may occur over short distances. In northern Algeria, for example, the rocks of young mountains present alternating soft shales, marls and schists with hard calcareous sandstones (Roose 1993:137). Maps of the superficial geology of an area will indicate the potential variety of soil parent materials.

In steeplands, however, the rocks underlying the existing soil may often not be the soil´s parent material if the rocks have been overlain by eroded sediments, landslide debris or other detritus derived from uphill sources, in which the soils have developed.

The texture (clay, silt, sand and gravel content) of a soil derives from that of the true parent material, and has a major effect on the structure and drainage of the horizons below the topsoil. In the uppermost horizons the effects of texture and structure are generally modified to a greater or lesser extent by the presence of higher organic matter contents (Shaxson 1992b:433; Shaxson 1993a:249).

Soil types: In steeplands in the tropics the main soil types are generally Andosols/Andepts[1] (derived from volcanic ashes), Leptosols/Fragmental or skeletal soils (soils of high gravel content), Regosols/Psamments (young soils developed in unconsolidated materials) and Cambisols/Inceptisols (soils with a cambic B horizon) (FAO 1991:13-36). Soils are usually extremely variable, due to site differences in geology, geomorphology, vegetation and rainfall (Cassel and Lal 1992:63).

Because of natural and accelerated (due to human actions) erosion processes, the soils on slopes are often young, either because the topmost layers are continually being lost, or because they are continually being overlain by the deposition of materials moving down slope. In neither case can distinct horizons become well-developed before further loss or gain of materials occurs to change the situation.

Effective depth: The effective rooting depth of soil in steeplands may also vary greatly. On slopes formed on landslide material or by 'soil creep', soils may be deep though of very mixed materials. On volcanic ashes they may be very deep and very uniform to depth (as in many Andosols/Andepts), even though the slopes are steep. But, particularly on upper slopes formed in other geological materials, from which the weathered products may have slid or crept down slope, soils may be very shallow due to continuous erosion of the weathered materials, or due to the formation of high gravel contents as in Leptosols/Fragmental or skeletal soils. On slopes at high altitudes, weathering processes may be very slow, which, coupled with geological erosion by wind and water, may also result in shallow soils (FAO 1991:17).

Soil depth represents a balance between the rate of soil formation and rate at which materials are lost from the surface (Hallsworth, 1987). Soil may be formed not only from the bottom upwards by very slow processes of rock weathering, but also from the top downwards by the more rapid processes of organic matter additions, humification and possibly tillage (Shaxson 1981a; Hall *et al.* 1982:357). In most steep land situations where there is inadequate land management, the rate of soil loss is much greater than the rate of soil formation.

Soil fertility: The 'fertility' of soils in steeplands may also be very variable, depending on the parent material which influences soil texture and clay mineral type, and on the type of vegetation/ land use and climate which will affect the organic matter content, nutrient additions and losses.

---

[1]   Throughout this Bulletin soil classification is presented according to both the FAO Legend (given as the first name) and the US Soil Taxonomy (second name).

vegetation/land use and climate which will affect the organic matter content, nutrient additions and losses.

VEGETATION ON STEEPLANDS

In areas of sparse population, some steeplands may still remain almost untouched and covered with climax native vegetation which is in a state of self-adjusting equilibrium with the local ecosystem. Some areas may have been significantly disturbed in the past, but have subsequently been left untouched for so long that an apparently 'indigenous vegetation' has re-established itself (Hallsworth 1987:6).

Few areas, however, now remain totally undisturbed, as most have been entered by people, at least to harvest forest products, to hunt wild animals, or to cut fodder for domestic livestock. If such harvesting has been so slight and/or so infrequent that it has not exceeded the vegetation's capacity to recover, the species density and composition may be little changed, and the soil little affected.

Chapter 3

Problems of farming steeplands

OCCUPATION OF STEEPLANDS

At least ten percent of the world's population, and a larger percentage of the world's poor, live in steeplands that are often associated with hills and mountains.

There are many reasons why people live, and sometimes prefer to live, in steepland regions:

- More agreeable temperatures and healthier living conditions than in nearby lowlands (e.g. Uluguru Mountains, Tanzania).

- A long history of living in the region (e.g. Andean chain of mountains).

- Insufficient available land in the lowlands (e.g. Philippines).

- Restricted availability of flatlands; by living on steep rocky slopes, all the good-quality arable land in the flatlands can be used for crops (e.g. parts of Lesotho).

- They or their forebears had to flee to areas of steeplands to find refuge from enemies (e.g. in the Dedza Highlands of Central Malawi).

- Soils on steep slopes (Cambisols/Inceptisols) may be more productive than the severely leached soils (Acrisols/Ultisols) on adjacent upland plateau areas (e.g. Papua New Guinea uplands).

Notwithstanding these perceived advantages, a farming family's situation on steeplands, particularly in mountainous areas, is often not only one of isolation, political marginalization and poor economic development, but also one of strong social and cultural cohesion (Douglas 1994a:1, 3-4).

USES OF STEEPLANDS

In extensive mountainous areas, there exists a vast range of combinations of agro-ecological environments and socio-cultural conditions, such that systems of land use and management can vary widely over relatively short distances. In smaller isolated steepland areas, the type of land use may be more closely related to the land use practised in the surrounding flatter lands, e.g. the same type of arable agriculture (as in some densely populated parts of southern Malawi), or it may differ from and complement the land use practised in the flatter lands, e.g. livestock grazing on steeplands to complement arable agriculture in the adjacent flatlands (as in parts of Lesotho).

Sequential systems of land use

As population pressures increase, there is often a change in the type of land use practised in steeplands:

- In the first phase, corresponding to extremely low population pressures, there may be non-destructive farming practices, such as the extraction of forest products (timber, fodder, honey etc). Provided the harvesting of natural products is in balance with the natural levels of production and regeneration, there should be no significant changes in either vegetation density, species composition or soil quality. Non-destructive farming is practised by some herders who migrate annually with their livestock between lowlands and highlands, using whatever fodder they can find on the way. This lifestyle is particularly common in semi-arid and arid regions.

- In the second phase when settlement occurs, shifting cultivation systems, often described as 'slash and burn' agriculture, are practised. These systems are more readily discernible on steeplands and have acquired a bad reputation, partly because they appear 'untidy' to many people. The key feature for successful shifting cultivation is the length of the regenerative 'bush-fallow' period, which should be sufficiently long to fully restore both the vegetation and soil physical and chemical fertility before the cycle is started again (Plate 2).

PLATE 2
Slash and burn farming, Nkhata
Bay, Malawi *[T.F. Shaxson]*

In the wet tropics of Asia the minimum period of bush fallow, under traditional farming systems, should be not less than 10 to 20 years (Siong 1990:20). Those areas where shifting cultivation provides a stable system of agriculture are characterized by low population densities, low living standards (Nicholson 1971:125), long fallow periods and in general by very low yields (Box 2). In some steepland areas, such as the Andean highlands, complex systems with long multiple cycles of cropping, grazing and fallow have been developed (Rist, 1992).

As population pressures rise, so the duration of the fallow period decreases. Too frequent cutting and burning of woody vegetation is likely to result in a partial or total lack of regeneration of many tree species. Without the presence of woody perennials that root deeply and bring up nutrients from depth to be returned to the soil surface in leaf litter, there is little possibility of soil structural regeneration or of adequate nutrient recycling. Thus, short fallows merely take land out of production with little or no benefit, and signify the commencement of soil and vegetation degradation.

• The third phase, in situations where population pressures are high, usually corresponds to the development of more or less continuous tillage with very short or no fallow breaks.

In these situations lands are often 'anthropized', i.e. modified by human actions, to make them more suitable for growing the preferred types of plants. This is accomplished by tillage, the addition of organic manures and mineral fertilizers, irrigation, drainage, and altering the shape of the land surface. The extent of Anthrosols is probably growing faster than any other soil group in the world, due to human impact on formerly undisturbed soils (FAO, 1991:14).

With clearing, fire and tillage, major changes take place in soil organic matter content, erodibility (as evidenced by 'K' factor changes), aeration, water-holding capacity, cation exchange capacity (CEC), pH, etc. (Willat 1993:203). In sloping areas where bench terraces have been constructed, soil modifications may be very significant, in terms of slope, water relations (e.g. waterlogging on rice terraces), depth (e.g. where terraces are built by cutting and filling), and organic matter content (e.g. where terraces are treated as favoured areas in view of the investment of much time and energy expended in their construction, and large amounts of organic manures are applied).

Resource-poor farmers, lacking the knowledge of appropriate management practices for steeplands, often impose unsuitable systems of land use which quickly result in land degradation. In annual cropping systems there is invariably a greatly increased exposure of soil to rainfall and insolation (Gibberd 1993:54), due to a lessening of the density, frequency or duration of surface cover. Soils become more exposed to rainfall intensities capable of initiating erosion, and to high temperatures that accelerate soil organic matter decomposition. The erodibility of soil may be significantly increased as a result of tilling annual crops causing pulverization and structural collapse. Ploughing can cause the downhill movement of significant amounts of soil by mechanical 'creep', which may be up to nine times greater than that moved by splash (Roose 1993:144). Even tillage with hand tools can result in pulverization of the topsoil and compaction of underlying layers. These effects may be just as catastrophic as those of a very severe natural event.

In SE Asia thousands of farmers accustomed to irrigated rice production have migrated into rainfed steepland areas where their inappropriate land-use practices have caused severe land degradation.

Instead of annual crops, the native vegetation may be replaced by planted pastures which are significantly different both in terms of the intensity of harvesting, frequency of passage of animals, the quality and quantity of organic matter produced, effective rooting depth and depth of moisture extraction. Some grasses such as *Cenchrus ciliaris* are able to extract moisture from as deep as 7 metres (Pereira 1973:159). Whether a grass vegetation is better or worse than the native vegetation in terms of rainwater infiltration is highly dependent on the management of the plant-animal grazing system.

Native vegetation may also be replaced by other perennial and semi-perennial crops such as cacao, oil palm, citrus, rubber, plantation-forest species, and sugar cane. Their effects on the soil will depend both on the nature of the plantings (whether in open mixed stands or in closed monoculture blocks), and how they are managed.

With respect to the quality and stability of the soil, as well as the quantity of useful and desired products which can be harvested from a given area of steepland, the effects of significantly changing the vegetation can be beneficial, neutral or harmful, depending on the management.

PROBLEMS OF FARMING STEEPLANDS

As rural populations expand, more and more people are likely to move into steeplands whether or not it is ecologically desirable, and whether or not the land-use systems they bring with them are sustainable in such environments (Hudson 1992:180). The settlement and farming of steeplands frequently brings a wide range of problems, social, economic and environmental, many of which are intimately interrelated.

Socio-economic problems

Many mountainous steeplands are remote with problems of access, and the inhabitants experience many disadvantages due to communication problems (Douglas 1994a:3, 4, 10a). In these situations there is often a lack of all-weather roads, inadequate transport, and no reliable telephone links with the more developed towns and cities in the lowlands.

In steeplands with access problems, inputs, such as seeds of improved varieties and of 'new' crops, fertilizers and pesticides are generally very expensive, not available, or arrive too late. Agricultural extension, veterinary and other services are frequently non-existent, or at best sporadic, and farmers in steeplands are frequently unable to acquire essential up-to-date information on better farming practices, marketing and opportunities for diversification. The effects of these factors are to maintain low levels of production and to constrain farmers to continued poverty.

Diversification will not only generate additional income for the farmers, but will also reduce their vulnerability to market and price fluctuations, and to the impacts of extreme climatic events. Diversification and better marketing opportunities are often key factors that can enable farmers to break through the poverty barrier that surrounds them.

Problems of access also increase the difficulties experienced in the acquisition of credit. The fact that most steepland farmers do not own property that can serve as collateral may make them ineligible to receive credit from some institutions. Even if they are successful in obtaining credit, the high interest rates and short repayment periods often mean that they are forced to sell their produce at a time of surplus when prices are low.

The greater access problems of steeplands compared to flat areas frequently compel farmers to sell their produce locally because of the difficulties of transporting produce to higher-paying markets further afield. Sometimes the very low prices obtained at local markets may not even cover production costs.

In many areas, farmers´ associations are either non-existent or largely ineffective. This prevents farmers from acting together to increase their income by, for example, the bulk purchase of inputs at lower prices, and bulk selling of produce at higher prices. By working together as a group or association, farmers may also be able to avoid the use of intermediaries and so benefit from increased profit.

The lack of farmers´ associations is particularly common in resettlement areas, where people from different origins, sometimes of diverse cultural and possibly political backgrounds, are settled in the same locality. In El Salvador, for example, ex-combatants from the recently ended civil war, originating from different parts of the country, have been resettled in the same locality. This has created much distrust and resistance against the formation of farmers´ associations.

As populations increase, both on steeplands and elsewhere in the tropics, there is a tendency for farms to be subdivided between family members, so that average farm size declines. It has been estimated that the minimum area required per person for an adequate diet is from 0.1 to 0.2 ha with modest levels of inputs (Lal and Pierce 1991:2). This tendency becomes all the more serious in steepland areas where the soils are often of low productivity.

PLATE 3
Bench terraces, probably constructed long ago in a rocky landscape, were located and shaped to make best use of scarce rainwater for crop production. Nowadays, because of the migration of people out of the area to obtain higher incomes, there is no longer sufficient labour to maintain and repair the terraces, and so they could lose their effectiveness. Yemen *[J. Demanguez]*

Communities in some densely populated steepland areas may suffer from outward migration. This may occur because of the 'attraction' of higher incomes elsewhere, as for example in Yemen where people are leaving their farms in the upland areas to earn wages in the oil-rich Gulf States. As a result there has been neglect and consequent degradation of the ancient terraces they were formerly cultivating (Vogel 1988:324; 1987) (Plate 3).

Outward migration has also occurred because farm sizes have become too small to support the increased population. Although high populations increase the quantity of potential farm labour, the available food may become insufficient, or the families' cash needs may no longer be satisfied from farm incomes alone, so many men leave for the towns and cities in search of paid employment (Tamang 1993:16).

On the one hand, the generally low average yields achieved by resource-poor, small-scale, farmers provide a 'window of opportunity' for increasing production but, on the other hand, farm sizes are becoming so small that:

"...this window is closing rapidly. Agricultural production can only be intensified while family plots remain large enough to yield surpluses to pay for ... essential inputs ... The most urgently needed inputs in many such areas under population stress are fertilizers and seed of improved crop varieties which are capable of profitable response to such nutrients." (Pereira 1991:21, 22).

Many of the socio-economic problems such as those related to poor access cannot be adequately overcome in the short term, whereas other problems such as the lack, or weakness, of farmers´ associations can often be solved relatively quickly with beneficial results.

Political problems

Farmers living in steeplands generally have much less political power than those living in more accessible and favourable lowland areas. Consequently, they are seldom able to influence government decisions on issues such as agricultural prices, subsidies and tariffs which greatly influence farm income.

The poor access of those living in steeplands is one of the reasons why the elite of some countries, who usually live in the main cities and wield most of the political power, generally pay more attention to the more 'visible' urban populations.

Environmental problems

High population pressures of people and livestock invariably lead to the rapid destruction of natural vegetation through severe harvesting of wood for fuel or building, intensive foraging and grazing by livestock, and by accidental or deliberate fires. The loss of cover intensifies soil degradation, especially on steeply sloping lands, causing more erosion and runoff, fewer springs, reduced river flows, and increased sediment deposition in lakes and rivers. The net result is a lack of potable water, firewood and fodder close to populated areas. Consequently farm families spend more time collecting these necessities, at the expense of tending animals and crops, than would families living in less degraded areas in flatter lands.

Increased incidence of flooding, land-slides and gullying in steepland areas can also seriously deteriorate road conditions and accessibility.

Soil degradation: The problems confronting farmers cultivating steeplands are frequently exacerbated by the inappropriate farming systems practised which provoke land degradation (Pereira, personal communication, July 1994). This may be partly attributed to steepland farmers lacking good extension services, information and available inputs.

Many of the soil degradation problems in steeplands are basically more severe manifestations of the same problems that occur on gently sloping lands, but the overriding problem is that of soil erosion by water.

Soil loss: Although the rainfall on steeplands may be no more erosive than that on nearby lowlands, the resulting runoff and eroded sediments will move much more quickly on steep slopes, and will provoke more severe rill and gully erosion. This is due to the greater velocity of runoff, and to splashed soil particles moving a greater distance downhill than uphill [Plates 4 and 5].

PLATE 4
Large raindrops can throw soil particles as much as 50 cm into the air, as shown by dis-colouration of the paint on this wall splashed by soil *[T.F. Shaxson]*

On steep slopes runoff will be faster than on gentle slopes, and a surface-flow velocity of more than about 25 cm/sec is generally considered erosive for bare soils. Therefore the combination of rain falling at intensities greater than 25 mm/hr (generally considered the erosive threshold value) plus erosive runoff can be a powerful and detrimental combination on bare steeplands. Runoff that is sufficiently fast-flowing may "shave off" the entire topsoil layer (Lal, 1981).

A rule of thumb for estimating the amount of soil splashed downhill relative to that splashed uphill is '50% + % slope'. Thus, on a 20% slope, about 70% of the soil would be splashed downhill and only 30% uphill.

Splash erosion is most commonly observed on poorly-managed croplands, whereas gully erosion is more common on poorly-managed grazing lands.

The loss of surface soil by erosion may have more serious consequences on steeplands because (a) most of the available nutrients are concentrated in the surface layers and (b) many steepland soils are shallow and so the loss of soil represents a bigger proportional loss than from deep soils on gentler slopes (Shaxson *et al.*, 1989; Cassel and Lal, 1992:62).

PLATE 5
Raindrops have splashed clay particles and organic matter into the furrows, further than the sand particles. If runoff occurs, the clay and organic matter, being finer, would be transported more readily, and further, than the coarser fractions. Bunda, Malawi *[T.F. Shaxson]*

Soil physical deterioration: Land clearing, tillage and overgrazing may alter soil structure at and below the surface through compaction, pulverization and loss of soil organic matter, leading to a deterioration of soil structure. When soil organic matter falls below about 2%, there is a marked decrease in soil structural stability, and soils that are of low erodibility under natural conditions may then become unstable and erodible if poorly managed (Cassel and Lal 1992:68, 86; Stocking 1995a; 229).

However, it is not changes in soil structure, i.e. the arrangement of the soil's solid particles, *per se* that cause the problems in root development, water movement and retention, but changes in the size and continuity of the soil's pores (Box 2).

To reflect this change of emphasis from the solid constituents to the spaces, the term 'soil architecture' may be used which refers to the distribution of spaces and pores between the soil particles, instead of the term 'soil structure' (Box 3). The loss of pores and spaces from soil architecture, by collapse and by clogging, often referred to as compaction (Figure 1), is the most significant physical effect of poor soil management.

Runoff and soil moisture loss: Soil architectural stability (i.e. pore stability) in the top few millimetres of the surface is of critical importance, since this is the first determinant of the partition of rainwater between infiltration and runoff (Figure 2). Degradation of the surface architecture through loss of porosity, by interstitial sealing, compaction, pulverization or collapse, will reduce soil infiltration rates and augment the loss of rainfall as runoff (Plates 6, 7 and 8).

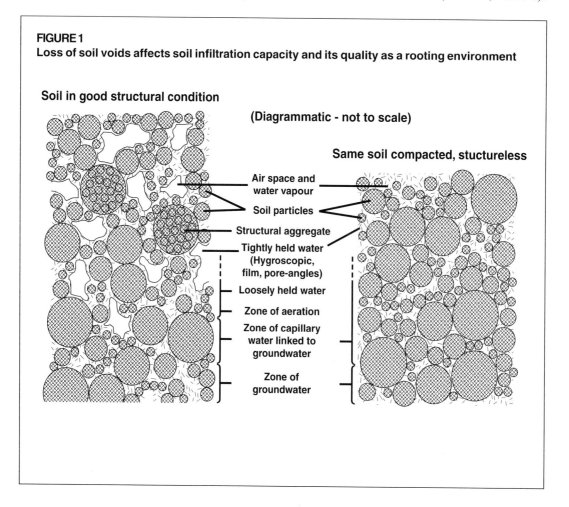

FIGURE 1
Loss of soil voids affects soil infiltration capacity and its quality as a rooting environment

Soil in good structural condition

(Diagrammatic - not to scale)

Same soil compacted, stuctureless

Air space and water vapour

Soil particles

Structural aggregate

Tightly held water (Hygroscopic, film, pore-angles)

Loosely held water

Zone of aeration

Zone of capillary water linked to groundwater

Zone of groundwater

BOX 2: SOIL PORES

"Soils in good physical condition are loose, moist and well aerated with well-connected macropores that allow roots to grow unimpeded. ... The conventional definition of soil structure as 'the arrangement of primary particles into aggregates' is not comprehensive enough to reflect all attributes that affect soil processes and plant growth. An edaphological definition must consider "those properties of soil that regulate and reflect a continuous array of various sizes of interconnected pores, their stability and durability, capacity to retain and transmit fluids, and ability to supply water and nutrients for supporting active root growth and development." (Cassel and Lal 1992:61, 65, 66)

BOX 3: SOIL ARCHITECTURE

In contrast with the term 'soil structure', 'soil architecture' emphasizes the spaces within and between the structural units. The spatial arrangement of solids and spaces within the soil affects (and is affected by) the functioning of soil-inhabiting organisms and plants' roots (Shaxson 1992b:433). As in a building, all the interesting happenings – gas exchange, root exploration, water movement and retention – occur in the spaces or voids of the architecture, not within the physical particles themselves.

FIGURE 2
Possible destinations of rainwater (FAO 1995:4)

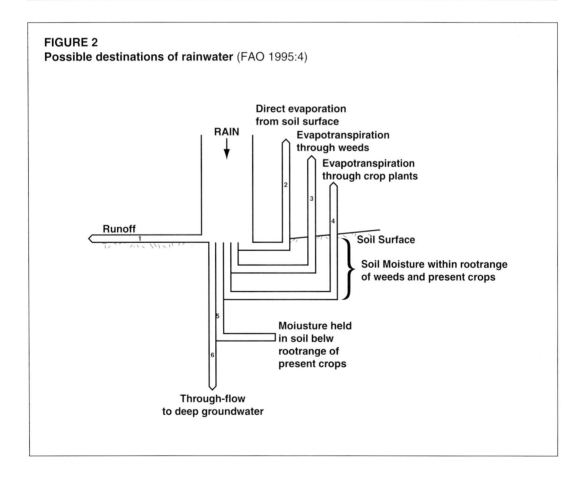

Figure 3 shows the contrast between the low infiltration rate of a severely degraded soil and the very high infiltration rate of the same non-degraded soil type under undisturbed forest. The effect of varying infiltration capacity of surface soil on the partition of rainfall between infiltration and runoff is diagrammatically shown in Figure 4.

PLATE 6
The underside (left) with good porosity, and upper side (right) of a surface crust with low porosity on a volcanic soil, caused by raindrop impacts. Tierra Blanca, Costa Rica *[T.F. Shaxson]*

PLATE 7

Runoff from a Vertisol on which the surface has become effectively sealed by raindrop action and swelling of the clay. Indore, India *[T.F. Shaxson]*

PLATE 8

Severe compaction of an unprotected soil surface caused solely by high-energy raindrops over two months. The experimental microplot for runoff measurements is about 2 x 1 m size. Mulanje, Malaw*i [T.F. Shaxson]*

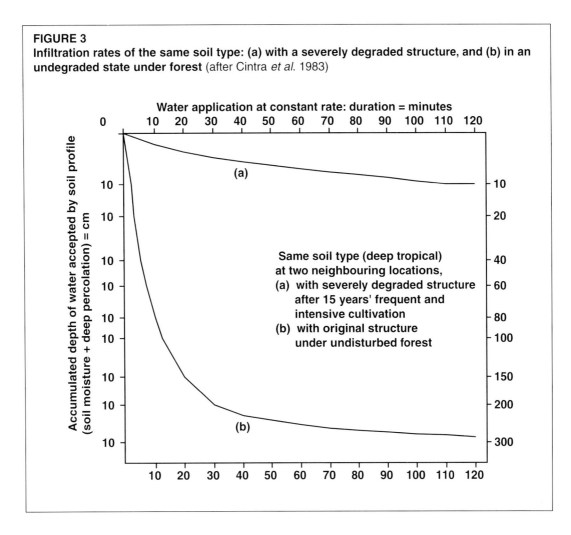

FIGURE 3
Infiltration rates of the same soil type: (a) with a severely degraded structure, and (b) in an undegraded state under forest (after Cintra *et al.* 1983)

Loss of water as runoff has been considered as the driving force responsible for soil erosion, but an alternative perception is that it represents the loss of a primary resource - water - and diminishes the proportion of the total seasonal rainfall that can be utilized for plant production.

Problems of water loss become more acute on steeplands than on flatter lands because of the tendency for runoff to increase with slope steepness. This problem is further aggravated in steeplands by the frequent occurrence of soils with low available water retention capacities due to their shallow and often stony nature.

The critical importance of the effect of soil cover on infiltration rate and hence on potentially crop-available water is shown by data in Table 1. In the presence of a good ground cover, represented by 'mulching', infiltration rate is significantly higher than from 'bare fallow'.

The loss of rainfall by runoff may induce more frequent and more intense water stress in plants than would have been the case if all the rainfall had infiltrated and become % 'effective'.

Successive periods of water stress during the life-cycle of plants result in a cumulative lessening of the potential final yield (Figure 5). However, plants with deeper roots, able to explore a larger volume of soil, will generally suffer less frequent and less damaging stresses than those with shallower roots.

TABLE 1
Effects of mulch on soil properties and rainfall acceptance (derived from Bonsu and Obeng 1979)

Practice	Bulk density mg/m3	Porosity %	Infiltration %
Bare fallow	1.72	35.1	56.1
Zero tillage	1.66	37.4	93.1
Mulching	1.55	41.5	97.1

FIGURE 4
Infiltration rate at the soil surface determines how rainfall is partitioned between infiltration and runoff

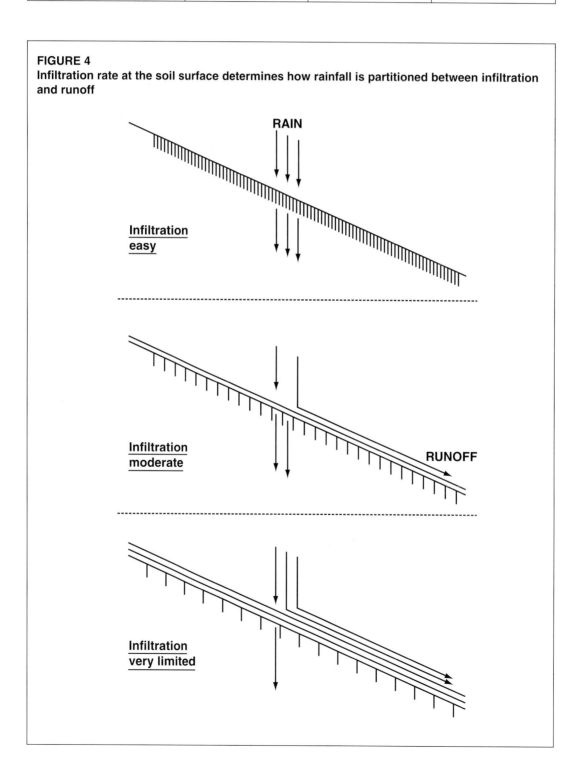

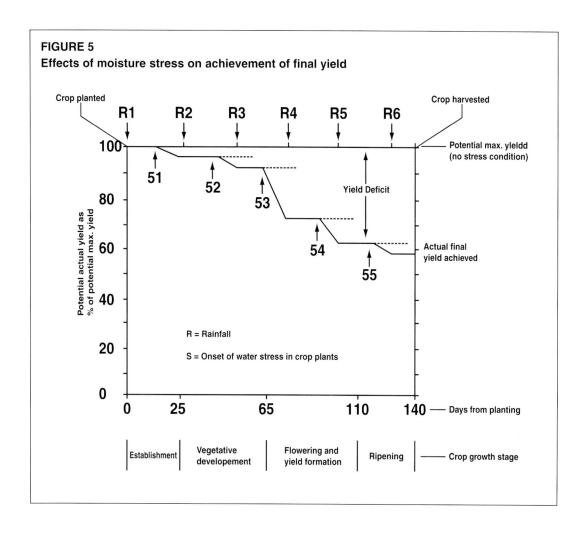

FIGURE 5
Effects of moisture stress on achievement of final yield

Reduction in effective rooting depth: Changes in soil architecture caused by land clearing, tillage or overgrazing may bring about soil densification in the root zone by compaction or structural collapse. If densification is very severe it may diminish the percolation rate of rainwater that has infiltrated, and so limit the exchange of O_2 for CO_2 in the root zone. More commonly, densification is only sufficiently severe to hinder the expansion, growth and functioning of roots due to increased physical resistance (Plate 9). Where pore diameters are less than about 0.1 mm, most roots can only penetrate by forcing their way through the soil pores. Thus, the combination of very small pores and high bulk densities may severely restrict root growth. The critical bulk density at which root growth begins to be limited depends on texture. One trial showed that roots of maize, soybean, wheat and beans were not affected in a sandy loam soil as compared with a clay soil. In a clay soil with 70% clay, bean roots were severely hindered at 1.1 gm/cm³, maize roots at 1.2 gm/cm³, and wheat at 1.3 g/cm³ (Cassel and Lal 1992:74).

Nutrient losses: Erosion and runoff carry away not only soil particles and organic matter with their associated nutrients, but may also wash away fertilizers and other agrochemicals. Losses of soil nutrients are accentuated by splash erosion and runoff due to the differential removal of the lighter and smaller, more nutrient-rich, clay particles and organic matter. Eroded soil materials may be 4-5 times richer in plant nutrients than the soil left behind. In addition, burning, tillage and harvesting also affect the flow of plant nutrients through the land system (Figure 6).

PLATE 9
Roots of soybean distorted by being unable to penetrate a tillage-induced subsurface 'pan'. Ponta Porao, Brazil *[T.F. Shaxson]*

TABLE 2
Representative elemental removal in various food crops

Crop	Yield t/ha	Element kg/ha				
		N	P	K	Ca	Mg
Corn (maize) grain	7	128	20	37	14	11
Rice (grain)	8	106	32	20	4	1
Wheat (grain)	5	80	22	20	2.5	8
Sorghum (grain)	8	135	10	27	16	9.6
Cassava (roots)	16	64	21	100	41	21
Potatoes (roots)	22	120	20	166	40	26
Bananas (bunch)	10	19	2	54	23	30

Source: Buol and Eswaran 1993:24, after Sanchez 1976.

Typical quantities of nutrients lost in the harvests of different crops are presented in Table 2. Long-term harvesting and removal of crops can result in nutrient impoverishment and declining soil fertility, if the rate of nutrient removal is not balanced by the rate of release of nutrients from soil minerals, by the application of organic or inorganic fertilizer materials, or by bush fallows. This is illustrated in Table 3 where continuous cropping without adequate fertilizer additions resulted in serious reductions in yields. It is tempting to think that yield decline over time is solely due to erosional loss of topsoil. But the extent of the compounded problems of disease, weeds and pests should not be overlooked.

"In Tanzania, Moberg (1972) compared soil analytical data from two eroded and two non-eroded profiles. He found that erosion had the effect of lowering pH to the extent that aluminium toxicity was commencing; organic carbon and nitrogen levels were halved in the top 15 cm; zinc and phosphorus were depleted to deficiency levels; and calcium and magnesium were considerably lowered. Fertility appeared to be affected to at least 150 cm in the profile, and Moberg suggests that this is brought about by the absence of roots in the eroded profile which prevents recycling. (Thus, more soil nutrients are leached out of the profile.) This is one of the few studies that highlights the crucial interrelationship between erosion and loss of productivity with explanations for the linkage." (Stocking 1995b:6).

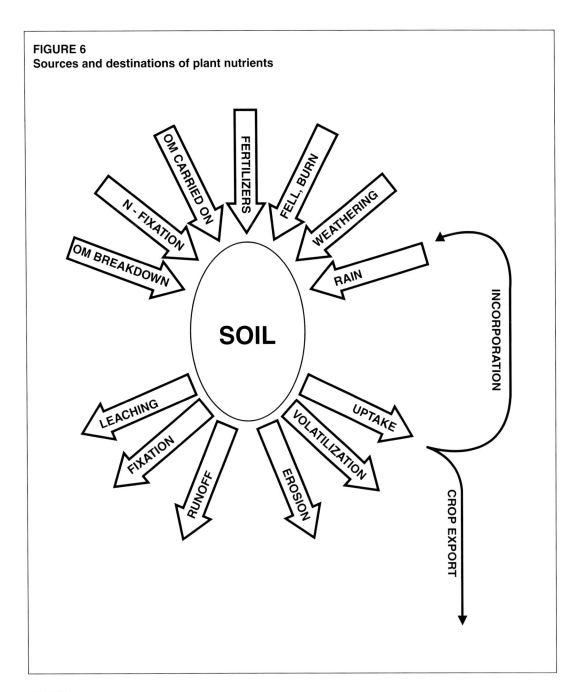

FIGURE 6
Sources and destinations of plant nutrients

TABLE 3
Effect of continuous cropping in Nigeria

5-year cropping period	Groundnuts (kernels) kg/ha	Millet (grain) kg/ha	Sorghum (grain) kg/ha
1931-1935	1015	920	540
1936-1940	785	455	330
1941-1945	700	320	105
1946-1950	320	545	90
1951-1955	510	300	discontinued

A more recent initiative to assist researchers in more than eleven countries to investigate these relationships in greater detail is being funded by FAO (Stocking and Sanders 1992:102-108).

Chapter 4

Past approaches to steepland problems

Past approaches at stabilizing agriculture on steeplands have been based on a logical sequence of thoughts and actions.

Governments have generally assumed responsibility as custodians of the land for the good of present and future generations, with a view to avoiding misuse and degradation of what is often the prime natural resource of many nations. It has therefore made sense to prepare strategic plans for the conservation and development of land on the basis of land-use capability and land-use hazards (Douglas 1994b:66; Hudson 1987:19; 1992:52).

In almost all land-use capability classification systems, land steeper than about 12-15% has been classified as 'non-arable' and too hazardous for the growing of annual crops. For such lands, the most common recommendations have been perennial crops, pastures, forests and the maintenance of natural vegetation (FAO 1983; Shaxson *et al.* 1977).

It has been widely observed that few signs of erosion occur under relatively undisturbed native forest; on the other hand, where such forest has been cleared for agriculture, and especially on steep slopes, subsoil exposure due to erosion has occurred within a short time (Plates 10 and 11). These observations have frequently resulted in recommendations that cleared steeply sloping lands should be re-forested to prevent further erosion. Laws have been widely enacted to give legal protection to both native and planted forests in an attempt to exclude agriculture from these steeplands.

Where slash-and-burn agriculture has infiltrated into areas designated as 'non-arable', it has been condemned as a destructive form of land use, and if the farmers could not be removed, then soil conservation measures had to be introduced.

PLATE 10
Denuding the soil surface and loosening it by tillage results in erosion of the surface soil. The transported particles have been filtered out of the runoff by the dense grass of the pasture below. Gualaceo, Ecuador *[T.F. Shaxson]*

PLATE 11
The loss of vegetation followed by hand cultivation for many years has resulted in structural degradation and runoff which takes much of the loose soil with it. Zomba, Malawi *(T.F. Shaxson)*

Both the general condition of the farms of small-scale, resource-poor farmers, and their apparent aversion to adopting the government's soil conservation recommendations, reinforced the view (often implied, and sometimes explicitly voiced) that farmers in tropical regions are unconcerned about the value of land, and must therefore be 'brought into line' by education and possibly coercion.

Erosion has been portrayed as a prime cause of land degradation and as a negative force to be halted, as exemplified by phrases such as 'the battle against erosion', 'the cancer of erosion', 'combating erosion', etc. These sentiments prompted the establishment of various soil conservation entities, e.g. branches, ministries and departments, in order to provide specialist attention to the problems.

Runoff and erosion have been seen to be damaging soil productivity through the loss of water (thereby reducing soil moisture availability) and by the removal of soil physical, chemical and biological constituents. It has been widely assumed that runoff is the chief cause of land degradation, and so most efforts have been directed at how to control runoff (Doolette and Magrath 1990:36, 37), with the hope that by so doing erosion would also be controlled.

It has been commonly assumed that if soil erosion reduces crop yields, then soil conservation will raise yields, although some disagreement exists as to the validity of this assumption (Doolette and Magrath 1990:52, 71, 72). Nevertheless, soil and water conservation measures were believed to be necessary to halt runoff and erosion before yields could rise. Because the relationship between erosion and the decline in crop yields has not been clearly defined, it was assumed (hoped) that the costs of implementing physical conservation measures would be compensated by the economic benefits accruing over a relatively long period – often as much as 20-30 years (Hudson 1987:168). Despite these uncertainties, and technicians' inabilities to demonstrate (rather than merely infer) these monetary benefits, many people assumed that the logic and benefits of 'soil conservation' would be self-evident to farmers.

Since erosion is more serious on steeplands than on flatlands, and since efforts to control it have not been wholly successful, there has been a demand for research to devise new technologies that can be 'transferred' by conservation extensionists to the (apparently unconcerned) farmers.

PLATE 12
Physical conservation banks that have not been maintained are no longer conservation-effective. Maluti Mountains, Lesotho [*T.F. Shaxson*]

It is the desperate need to halt erosion that has motivated the design of projects focusing on soil conservation. In the last 40 years or so, many soil conservation projects have been instigated and funded by donor agencies, with the aim of 'intervening' so that farmers will implement recommended soil conservation measures (Anderson and Thampapillai 1990:1, 2). The tendency has been to establish projects of the demonstrative intervention type: "Stand back and we'll show you how to do it." Huge amounts of money have been spent on this type of project, but results have been generally well below expectations (Doolette and Magrath 1990:35).

More recently there has been a growing understanding of farmers' viewpoints and needs, and of the role of soil improvement in 'combating erosion'. This has led to increasing efforts to develop more 'rational' farming and agroforestry systems for farmers (Norman and Douglas 1994; Tejwani 1994).

However, despite abundant demonstrations of physical and biological conservation measures, there has been little spontaneous or sustained adoption of these practices. For example, in Lesotho there has been almost no spontaneous spread of the narrow-based banks, terraces and grass/stone waterways that were promoted by the Government and its aid partners (Plate 12), even after 15 years or more of incentives provided by Food for Work programmes (Shaxson, pers. obs.).

In Ethiopia, huge efforts have been made to stem soil erosion; widespread areas have been covered with terraces, and millions of trees have been planted, but in spite of these efforts, there has been little acceptance or replication of these practices by farmers, and little maintenance (Herweg 1993:392; Tato 1992:435).

Rural communities are frequently unwilling to adopt particular soil management recommendations for improving productivity if the recommendations are likely to increase the risks of failure, or if they require even the smallest sacrifice of production in any year. Such consequences could jeopardize the very survival of the communities, as many of these people have no cash with which to buy necessary inputs, and exist at the 'threshold of poverty' (Laing and Ashby 1993:64, 73). Small-scale, resource-poor farmers in steeply sloping lands frequently exist in conditions of extreme poverty - economic, environmental and, in some cases, even social poverty, as in some settlement schemes.

ENCOURAGING SIGNS

Some conservati on-effective land-use and management practices have been accepted by farmers, and have spread spontaneously as a result of, or even in spite of, project-type interventions. These practices have generally been those that have improved the stability or profitability of production. They have usually been practices, such as the more effective utilization of organic materials, which show quick and substantial benefits, such as more stable or higher yields, or less labour (Roose *et al.* 1992; Pretty *et al.* 1998).

In some steeplands where market conditions have generated a demand for crops from which farmers can gain acceptable levels of income, considerable effort and money has been invested by governments and farmers on bench terracing and road con-struction to provide a sustainable physical basis for commercial production. Examples are found in Taiwan (Moldenhauer *et al.* 1991), and elsewhere (Plates 13 and 14).

There exist examples of formerly stable systems of land use in steeplands, such as in Nepal, Yemen, the Philippines and Peru, where techniques and structures, such as cross-slope terraces, have evolved over considerable periods of time when population growth was much slower than it is today. Evidently these considerable works of terracing (Plate 15) are the culmination of a repetitive process of experience-feedback-adjustment-maintenance, which served the purposes of the people who developed them.

PLATE 13
Contour bench terrace constructed specifically for growing fruit trees in response to market opportunities. Jadan, Ecuador *[T.F. Shaxson]*

PLATE 14
Young coffee on bench terraces. Puriscal, Costa Rica *[T.F. Shaxson]*

PLATE 15
Farming crops and trees on carefully maintained steep hillsides. Nepal *[J. Clark]*

Since population pressure on the land was probably less than it is today, it may be surmised that such systems were installed and managed <u>not</u> specifically for soil conservation, but probably for other reasons, such as water management which gave better crop growth and higher yields. Perhaps this provides a key to the better understanding of why some practices are readily accepted by farmers, i.e. not because they control erosion, but because they increase productivity! This is also consistent with the fact that farmers seldom place 'soil erosion' (in the way that technicians conceive it) high on their list of priorities.

Chapter 5

Changes in concept and approach for improved steepland management

PERCEPTIONS AND ASSUMPTIONS

When looking at eroding steep lands, it is very easy to accept the common belief that 'stopping erosion' is the solution to the problem, and that people must be informed about 'soil conservation'. However, this approach has not been conspicuously successful in the past, and other ways of looking at the problem may provide different insights.

The first step is to be prepared to acknowledge that it may be possible to see and interpret an object or situation in more than one way. Two simple examples are given in Figure 7 and Plate 16.

In Figure 7, it is possible to see some sort of trumpet or chalice by focussing on the black area, or the heads of two people facing each other by focussing on the white areas.

Plate 16 may be interpreted as showing how much runoff there must have been to cause so much gullying; alternatively it could be interpreted as an indication of how much of the rainfall did not infiltrate into the soil, was wasted as runoff, and so could not be used for plant growth.

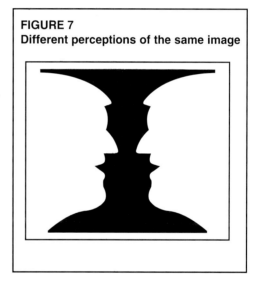

FIGURE 7
Different perceptions of the same image

Next, one should be prepared to acknowledge that another person – a farmer, for instance – may perceive a particular object or situation in a different way to one's own perception (Plate 17).

It is also necessary to beware of making untested assumptions, lest some decisions are made on misconceived beliefs, as illustrated by the examples in Box 4 and Plates 18 and 19.

More knowledge of the nature of a phenomenon and its processes can alter the perception of it, as in the case of 'what is the dominant process of soil erosion?' Contrary to the common perception that runoff causes most erosion, measurements showed that splash was the dominant process (Box 5).

PLATE 16
One perceives this as an area almost completely denuded by overgrazing; an alternative perception is that most of what little rain may fall in this area runs off the surface and cannot benefit the soil´s moisture reserves. The chances of vegetation re-establishing itself are correspondingly reduced. Kondoa/Irangi, Tanzania in 1961 *[T.F. Shaxson]*

PLATE 17
A field in Costa Rica had been laid out with physical works by field staff with the expectation of conserving soil. But when the farmer was asked what he liked about the layout, he replied: 'It makes the process of applying irrigation water much easier'. Costa Rica *[T.F. Shaxson]*

BOX 4: HOW ONE'S OWN UNTESTED ASSUMPTIONS CAN MISLEAD

"In an area of Lesotho, I had observed on many occasions that farmers were harnessing as many as six oxen to a single-furrow plough in order to till soils which I had earlier surveyed and found to be reasonably friable. I noted that the yokes were merely bluegum poles with pegs through them, and looked particularly uncomfortable for the oxen by comparison with the collars and pads provided for the horses and donkeys which pulled carts around the nearby town. I assumed that, because of the uncomfortable yokes, each ox was not exerting full power, and that therefore several oxen were needed. I reasoned that, if more comfortable yokes could be obtained, fewer oxen would be needed per plough, and many would thus be released for work with a greater number of ploughs so that more land could be prepared at the critical time just before planting. I thought this might be a useful sub-project to develop. But a farmer to whom I expressed this idea laughed and told me I had got it all wrong. He said: "If I have six oxen and it only needs two to pull the plough, two will be employed. The other four will be wandering around the village making mischief, unless I employ someone to keep an eye on them. So to keep them all under control without extra cost, I hitch them to the plough too."

"Three lessons:

1. A farmer's perceptions and reasoning may differ from yours, and they may be more valid than yours in the circumstances.
2. Examine the validity of your own assumptions.
3. Discuss matters with farmers at an early stage."

(from Shaxson 1993b:121)

PLATE 18
Six oxen hitched to a single-furrow plough on soil that would appear easy to cultivate. Note the uncomfortable-looking yoke on the rear pair of oxen. Mafeteng, Lesotho *[T.F. Shaxson]*

PLATE 19
By comparison, the ponies at the local market had harnesses that appeared more comfortable in which to work. Mafeteng, Lesotho *[T.F. Shaxson]*

BOX 5: RAINSPLASH VS. RUNOFF

"We found flowing water tore loose less than 10 percent of the soil moved from our fields; in most cases, it was less than 5 percent. In other words we found that about 95 percent of the soil eroding from our fields is splashed by falling raindrops. ... We know now why our fields continue to erode after being 'terraced'. Thinking our enemy attacked from the sides, we built walls when we should have built roofs. Our main attack was from above, so we were fully exposed to our attackers but didn't know it."   (Stallings 1957:34)

BOX 6: VERTICAL THINKING

"Lateral thinking is made necessary by the limitations of vertical thinking. The terms 'lateral' and 'vertical' were suggested by the following considerations.

"It is not possible to dig a hole in a different place by digging the same hole deeper. Logic is the tool which is used to dig holes bigger and deeper, to make them altogether better holes. But if the hole is in the wrong place, then no amount of improvement is going to put it in the right place. No matter how obvious this may seem to every digger, it is still easier to go on digging in the same hole than to start all over again in a new place. Vertical thinking is digging the same hole deeper; lateral thinking is trying again elsewhere ..."

(De Bono 1967:22)

Lateral thinking is useful for escaping from the confines of conventional thinking (Box 6), and for developing new concepts that can be used to change the approaches used in overcoming the problems of farming steeplands.

Examples of the new concepts that better address the problems confronting small-scale resource-poor farmers on steeplands are presented.

CONCEPT A

Old concept: **Runoff and erosion are the primary causes of land degradation.**

New concept: **Runoff and erosion are the consequences of certain features of land degradation, viz the loss of soil cover and loss of soil porosity.**

Note: This concept relates to accelerated soil erosion processes induced by human actions, not to naturally occurring geological processes.

Comments

Erosion and runoff have commonly been regarded as the causes of land degradation, and the causes of soil erosion and runoff are generally cited as 'deforestation', 'overgrazing' and 'over-cultivation'. So the general advice given to farmers has tended to be: 'Don't cut down the forests', 'Don't overgraze' and 'Reduce your cultivations'.

TABLE 4
Common consequences of deforestation, overgrazing and over cultivation

Macro ⇨ ⇨ Micro ⇩ ⇩	Deforestation	Overgrazing	Over cultivation
Loss of effective ground cover Loss of organic materials and activity Loss of soil porosity	* * *	* * *	* * *

Even with prohibitive legislation, these instructions have seldom been willingly obeyed by small-scale resource-poor farmers (or indeed by any other farmers). However, it is very revealing when one realizes that these three actions have several consequences in common – loss of ground cover, loss of soil organic matter, reduced biological activity, and loss of soil porosity (Table 4). It is these features which represent land degradation, and which are caused by inappropriate land use or by poor land management.

Runoff and erosion are therefore the consequences of the following features:

i. a diminution of cover and hence reduced effectiveness in lessening the impacts of erosive raindrops;
ii. a deterioration of the soil's architecture resulting in loss of porosity and reduced rainfall infiltration into the soil;
iii. a decrease in soil structural stability, often due to lower soil organic matter contents, and hence greater soil erodibility.

Erosion is thus not some sort of monster or cancer moving across the land to be 'battled', 'fought', 'combated', 'stopped', 'stemmed', etc. as is commonly suggested.

On the contrary the new perception is to appreciate that deterioration of soil quality is the chief problem, and that this can be minimized through more appropriate land use and better land management practices that improve ground cover, soil porosity and structural stability. It is therefore better to tackle these factors (cover, porosity and structural stability) than attempt to control the three apparent culprits head-on as has been done in the past.

PRINCIPLES FOR FIELDWORK

* Attend to improvements in ground cover, soil architecture and soil porosity simultaneously, to simulate as closely as possible the 'ideal forest-floor' conditions (see Concept 'B'). Improving the ground cover of a soil with poor architecture will not be sufficient – nor will improvements in soil architecture if not combined with greater soil cover. Combine the improvements to soil cover and soil architecture so as to create conditions that are 'infiltration-effective' and hence 'conservation-effective'.

* Emphasize rain-**fall**/soak-**in** relations more than rain-**fall**/run-**off** relations as in the past.

On slopes up to about 20%, a 40% ground cover has been found to reduce soil losses to about 10% of the losses that occur from bare soil (Shaxson *et al.*, 1989). However, the percentage cover needed to ensure low risks of erosion for a given soil type and cropping system would be expected to increase with increasing slope gradient. Evidence to support this hypothesis was found in El Salvador, where a 75% ground cover was required to ensure low erosion risks on slopes of 20-60% (Barber, 1998).

On very steep slopes, ground cover which has a high degree of contact with the soil surface, such as plant residues, stones and mulching materials (referred to as 'surface contact cover'), may be more important than ground cover with a negligible soil surface contact, such as the canopy of crops and weeds (referred to as 'canopy or aerial cover'). Surface contact cover is sufficiently close to the soil surface to be capable of reducing the velocity of overland flow which transports detached soil particles, as well as protecting surface aggregates from particle detachment by the direct impact of raindrops. In contrast canopy or aerial cover can only protect against the soil detachment process (Paningbatan *et al.*, 1995:201).

Illustrations

The provision of ground cover to protect soil surface aggregates from the force of rain drop impacts is critical for maintaining a soil's rainwater infiltration capacity (Plates 20, 21, 22 and 23).

Figure 8 shows how the presence of a ground cover maintains soil infiltration rate, how its removal and exposure of the soil leads to crust formation and a very rapid decline in infiltration rate, and how, once the crust is removed, the application of a cover again exerts its protective effect. A simple demonstration of the influence of cover and soil structure on rainfall infiltration and runoff, that is suitable for farmer field days, is presented in Annex 5.

PLATE 20
The force of a large drop of falling water is dissipated by being broken up into a multitude of droplets on hitting a plant leaf. (The next drop appears above as a vertical line, a function of the camera's shutter-speed.) *[T.F. Shaxson]*

PLATE 21
A high degree of cover over the soil is provided by a relatively dense planting of maize *[T.F. Shaxson]*

PLATE 22
Next to Plate 21 was a poorer patch of maize of the same age. Note the sparser cover provided to the soil. Both photographed from the same place *[T.F. Shaxson]*

PLATE 23
A cropping system of maize undersown with Mucuna provides year-round protection to the soil surface and adds N-rich organic matter as well. Santa Catarina, Brazil *[T.F. Shaxson]*

FIGURE 8
Effect of protective cover on infiltration (Duley 1939, quoted in Schwab *et al.* 1966:68)

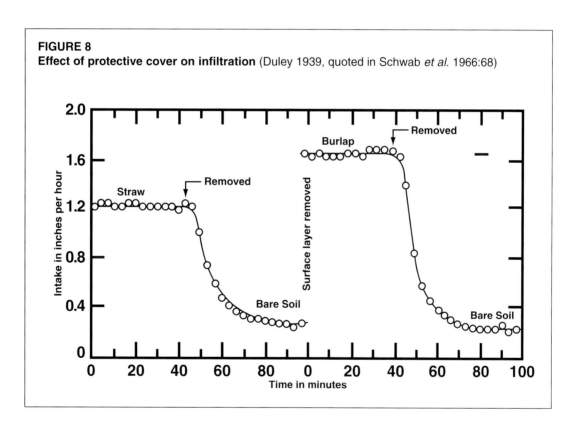

After land clearing and the commencement of tillage, the structural stability of many soils, and especially sandy soils, decreases very rapidly (Plates 24 and 25). This is probably related to the very high initial decomposition rates of the soil organic matter. As structural stability decreases, so soil erodibility will increase.

PLATE 24
This very sandy soil has been tilled for the first time after being cleared from *Brachystegia* woodland. While its architecture shows some structure, it is evidently very fragile

PLATE 25
The same soil type, on the adjacent plot, which has been tilled two years in succession. The original architecture has virtually disappeared, leaving it almost without structure. This is comparable to demolishing a building but leaving the pile of bricks in place. The soil is no longer as healthy because the useful voids in the soil (equivalent to rooms in the building) have been lost. Mubangwe, Malawi *[T.F. Shaxson]*

CONCEPT B

Old concept: **Stability of land productivity on steeplands is achieved by the installation of soil conservation structures.**

New concept: **Stability of land productivity on steeplands is achieved by introducing land management practices that aim to recreate, in as far as is practically feasible, 'ideal forest-floor' conditions.**

Comments

There has been insufficient success to date in stabilizing the productivity of inhabited steeplands, despite expenditure of much money, effort and legislation on the implementation of soil conservation structures. It would appear that continuing along the same lines of attack as before is unlikely to yield any better results than in the past, even if more money is applied, because farmers are generally unwilling to apply current recommendations, and conventional soil conservation structures are not adequate to stabilize productivity.

A new approach to stabilizing and sustaining the productivity of steeplands depends on a clearer and more widespread understanding of land and the ecology of land, which can lead us to change what has been the 'conventional wisdom' in soil and water conservation.

It is helpful to perceive land in terms of its various dimensions, viz. the three dimensions that define its external topography, the three that define its internal structure or architecture, the dimension of time along which changes occur, and the emotional dimension representing the strong affinity that people develop for the land they occupy (Shaxson, 1994a).

The ecology of land is determined by the particular combination of factors which constitute land, i.e. geology, topography, hydrology, soil, vegetation, macro-, meso- and micro-organisms, and by the continuous interactions of these factors under the influence of climate, gravity and people´s actions over time. As these factors change, so the ecology of the land will change.

Relative ecological stability of land is achieved by the presence of a permanent and continuous vegetative cover and by organic, biological processes which maintain the upper soil layers in continuously plant-friendly conditions. Reductions in plant cover or a decrease of organic activity will have the effect of diminishing the continual regeneration of plant-friendly topsoil. Top soil may then become degraded, or may be rapidly removed thereby exposing underlying layers that are less favourable for plant roots.

A suitable criterion of the 'stability' of a landscape and its components is the condition of relative equilibrium achieved in an undisturbed forest. The ideal situation, exemplified by a dense undisturbed forest, is land with minimum runoff and little erosion; relatively smooth variations in streamflow during the year; clean water in the streams; abundant ground litter beneath a high canopy; large quantities of biomass; much soil biological activity; good porosity; plenty of soil moisture; deep roots; and an almost closed cycle of nutrient transfers.

In these conditions rainwater is recycled for much of the year via evapotranspiration, depleting soil moisture reserves to the full depth of the trees' root systems, and thus continuously emptying pore-spaces within the profile which once again become available for moisture storage from the next rainfall event.

In terms of streamflow, the soil body under forest conditions acts as a buffer for water, accepting irregular inputs of rainfall over a few minutes, hours or days. Moisture which is stored at plant-available tensions may be returned to the atmosphere via evapotranspiration, while that which percolates past the root zone eventually reaches the subsurface water, emerging at seeps and springs, or goes further down to deep groundwater. Because of the buffering and filtering effects of the vegetation and soil, water which arrives as intense rainfall at the surface departs later as clear relatively persistent streamflow which is smoothed out over time due to its passage through the soil and underlying materials.

Even in areas of extremely high rainfall of ca. 10 000 mm/yr, as in parts of SE Asia, native forests are stable provided there are no very severe climatic events (Woruba 1990:41; Willatt 1993:196). Such conditions may also be achieved under well-managed tree-crops such as cacao, or oil palm (Maene *et al.* 1979).

Forest soils are typically characterized by high infiltration rates which allow for rapid and full recharge of soil moisture, and the transfer of excess water to groundwater reserves if these have been depleted. It has been reported that even in a mature closed-canopy 'miombo' woodland of seasonally-dry Zambia, the infiltration rate of un-burnt woodland research plots, with a spongy organic surface litter, was consistently above 30 cm/hr (Cheatle, pers. comm. 1992). Under high dense deciduous forest on a red latosol (Ferralsol) in southern Brazil, infiltration rates still exceeded 200 cm/hr after 30 minutes' continued application of simulated rainfall at a constant rate (Cintra *et al.* 1983). Under oil palm in Malaysia, initial and steady infiltration rates of 150 cm/hr and 50 cm/hr have been recorded (Maene *et al.* 1979:212), and rates of 13-24 cm/hr are reported for undisturbed soils in Puerto Rico (Doolette and Magrath 1990:191).

Since maximum rainfall intensities seldom exceed about 180 mm (18 cm)/hr for a few minutes in any place in the tropics, it can be appreciated why forest areas in good condition are seldom subject to severe runoff, except in the case of unusually severe rainfall events. With little runoff, there is also little soil movement.

These criteria indicate how much rainwater land can absorb, and how little soil is moved, in undisturbed conditions, which should also be the aim of well-managed agricultural land.

The stability of soil in the face of variations in climatic conditions – particularly severe events of rain and wind – is significantly affected by the organic materials and biological processes within, and immediately above, the soil. Under conditions of excellent cover, even the most fragile soils with very weak architectural units, when undisturbed, can maintain their infiltration capacities and porosity (Cassel and Lal 1992:79).

The quantity of crop residues required to maintain a continuous and effective protective cover on the ground will depend on the decomposition rate of the residues, which in turn will be determined by the chemical and physical nature of the residues and by climatic (temperature and rainfall) conditions.

Considerable 'churning' of the soil results from meso-organisms' activities (worms, termites etc.). In some situations, termites may bring as much as 5 tons/ha of subsurface materials to the surface each year (Stocking 1996). In ideal conditions in Nigeria, worms may produce at the surface up to 100 t/ha/year of casts, which have passed through their guts (Cassel and Lal 1992:67). These churning activities are responsible for the incorporation of organic materials into the soil and are vital for the creation of macro-pores that contribute to the soil´s porosity.

Thus, in 'ideal forest-floor' conditions, the high infiltration capacity of the soil is maintained through the combination of:

• stable porous architectural conditions,
• high biological activity, and
• a protective cover of litter and vegetation over the soil surface.

Principles for fieldwork

Place emphasis on those management practices that will maximize the production of biomass, maximize the percentage cover over the soil surface and its continuity with time, promote biological activity, and optimize the potential for nutrient recycling.

Wherever possible utilize crops, varieties, crop rotations and planting densities that produce large quantities of biomass. Try to ensure that residues are not removed from the field, burned, or grazed.

Introduce cover crops or multipurpose trees to augment the production of biomass and litter.

Minimize disturbance of the soil by practising zero tillage or, if tillage is unavoidable, use tined implements that leave a large proportion of crop residues on the soil surface.

Illustrations

Examples are presented of the creation of a high ground cover through the retention of litter (Plate 26), undersowing with a cover crop (Plate 27), and by zero tillage (Plate 28).

PLATE 26
Under eucalyptus trees the protection of the soil surface may be excellent if the litter is not removed, grazed or burned. Tupassi, Brazil *[T.F. Shaxson]*

Concept C

Old concept: **Declining yields are related to the quantities of soil, water and nutrients lost in the erosion process, therefore the primary cause for concern is the quantity of soil particles and water lost.**

New concept: **Declining yields are related to the quality of soil that remains after erosion, therefore the primary concern is how to restore the productivity of the existing soil through improved management.**

Comments

This new concept has been recognized for some time (Rosenberry *et al.* 1980; El-Swaify and Dangler 1982:12-14), but has not yet received the urgent attention it deserves.

If technicians are unable to demonstrate unequivocal relationships between the loss of soil and loss of yield, it is unreasonable to expect farmers to perceive such a relationship, or to expect farmers to accept our claims that erosion is a serious problem and to willingly adopt our recommendations for 'implementing soil conservation'.

A key point to consider is that 'soil productivity' resides not only in the soil´s chemical nutrients (of which a significant proportion may be lost by erosion), but also in the chemical, physical and biological components of the soil and their interactions.

PLATE 27
Excellent conditions of the forest floor may be simulated by mixtures of planted species, here oil palm undersown with a creeping legume. Anki Mabela, Fiji *[Natural Resources Institute]*

PLATE 28
The farmer found the retention of wheat-straw residues on the surface with zero tillage so effective in promoting water absorption (and preventing soil movement) that the broad-based banks previously instal-led to prevent runoff were no longer necessary. To facilitate management and limit wear and tear on his machinery, he now plants in straight lines across the (diminished-size) banks, without incurring runoff or erosion. **Campos Novos, Brasil** *[T.F. Shaxson]*

PLATE 29
Here the upper colluvial soil layer appears to be richer than the lighter-coloured subsoil materials, such that erosion would ultimately expose significantly less-fertile soil materials. **Tarija, Bolivia** *[T.F. Shaxson]*

The effect of erosion is to expose layers that were formerly below the surface, and which then become the new rooting environment for subsequent crops, pastures or trees. In most situations, the newly-exposed soil presents a poorer rooting environment (Plate 29), and when it becomes exposed to rainfall, intense insolation and strong winds, the plants suffer from increased temperature, more nutrient and moisture stress, and greater impedance to root growth (Williams 1990). Post-erosion yields will then be poorer than previously, because the plants are growing in a soil of inferior quality as manifest by:

- reduced depth for rooting and moisture retention;
- reduced quantities of nutrients and less available nutrients;
- less soil organic matter and lower biological activity;
- poorer soil architecture, i.e. lower porosity, and hence decreased infiltration and percolation rates, reduced available water capacity, less plant-available water, and slower gas exchange rates.

It is these soil qualities, presented above, which need to be improved to halt the declining yields and restore productivity.

Principles for fieldwork

Always aim to restore and improve the quality and depth of the remaining exposed soil as quickly as possible, paying particular attention to soil architecture, soil organic matter, and biological activity.

Introduce practices that will maximize rainfall infiltration. Since soil moisture is regularly depleted by evapotranspiration, always maintain soil cover and soil architecture in conditions that will permit the maximum possible replenishment of soil moisture from the next rainfall. Ensure as far as possible that the soil surface is always covered with crop residues, mulches, composts, crops or cover crops.

Promote the formation of humic, fertile soil developing from the top downwards, and enhance biological activity (Stallings 1957:116; Shaxson 1981b:357; Pretty 1994:1) by the use of organic manures and composts.

The processes of compost and manure formation both involve the passing of vegetation through the grinding-mills of animals' mouths (worms or livestock, respectively), and then the infection of the macerated vegetation with organisms that normally inhabit the animals´ guts, before excretion as worm-casts or manure, respectively. Heaps of vegetation residues for compost and manure production can be 'seeded' with earthworms to speed up the processes of organic transformations.

Optimize the amounts of organic materials occurring just above and on the soil surface, because of the many simultaneous benefits they provide:

- protection of the soil surface against the direct impact from raindrops;
- protection of lower layers of organic materials against insolation and oxidation, if undisturbed;
- moderation of soil-temperature fluctuations;

PLATE 30
A soil with high organic matter content is favoured by earthworms. They churn the soil as they burrow through it, incorporating organic materials. The burrows provide large channels for the downward movement of water and exchange of O_2 and CO_2 within the rooting zone. Burgay, Ecuador *[T.F. Shaxson]*

PLATE 31
'In-row tillage' system where crops are grown in double rows either side of narrow compost-filled trenches on the contour combined with mulching enhances water infiltration, raises soil productivity and achieves better weed control. Honduras *[J. Pretty]*

- provision of a low-concentration source of plant nutrients;
- provision of substrates for earthworms and other invertebrates that aerate and churn the soil, and incorporate organic materials (Plate 30);
- provision of substrates for micro-fauna and -flora which release nutrients and produce humic gums that improve structural stability;
- increases in soil cation exchange capacity (CEC), and hence an increased ability of soils to adsorb, retain, and allow slow release of, plant nutrients. Although organic matter in agricultural soils seldom exceeds 5% by weight, it often provides 40-50% of the soil´s total CEC (Flaig *et al.* 1977:7).

For direct enrichment of the root zone, incorporate the compost or manure (Plate 31).

Encourage resource-poor farmers, who are probably unable to purchase chemical fertilizers, to maximize the recycling of organic materials through the application of manures and composts, as this may be one of the few options available to them to maintain soil productivity and stability, even though the productivity of their soils is relatively low (Box 7).

BOX 7: COMPOST AND FERTILIZER

The advantages of compost

"They say it would be impossible to grow food as they do now without it and that it is good for the soil because:

- It makes the soil friable (loose, light and crumbly) and easier to plough because there are fewer clods.
- Excess water percolates through and moisture retention capacity is improved. Composting helps to regulate the amount of moisture in each plot according to the farmers' requirements.
- Allows air into the soil and makes it easier for the plant roots to breathe (improves aeration).
- Poor soil over time can be improved by application of compost.
- Reduces and remedies the negative qualities of chemical fertilizer.
- Reduces and remedies 'sourness' in soils.
- Provides nutrients for plant growth.

... The yield from up land [sic] from using compost and chemical fertilizer together is higher than when either is used separately. ... The farmers ... concentrate on finding options that simultaneously fulfil the increased demand for plant nutrient and maintain the structure of the soil."

Recognizing the need for mineral ('chemical') fertilizer

"The farmers' groups agreed that the rapid increase in population, division of the land into smaller plots through inheritance, and the subsequent need to grow more food have made it essential to increase crop yield. In order to do so they have intensified cultivation to a state where there is little or no time for fallow. They cannot sustain this intensity without chemical fertilizer. ... As far as the majority of the farmers are concerned its advantages are as follows:

- It is possible to cultivate the land throughout the year. This has enabled many small farmers [with very small plots] to grow sufficient food.
- Collecting materials, processing, transporting and applying compost require large amounts of labour. Applying chemical fertilizer reduces this requirement.
- It has made white soils productive which compost by itself could not accomplish.
- It increases yield in a short period of time. This is of particular interest to tenant farmers.
- It is easy to apply, one can learn by observation or simple instruction from other farmers.

"Disadvantages of chemical fertilizers ... are known to all the farmers but they still use it wherever possible. The drawbacks of using it on its own are:

- It makes the soil hard and dry and difficult to plough.
- It increases the labour requirement for ploughing and breaking clods.
- It reduces the moisture retention capability of the soil.
- Chemical fertilizer is not a real 'maal' [organic material which stabilizes, conserves or improves soil or soil fertility]. It is an agent that sucks the organic material out of the soil and makes it available to plants in a very short time. This process of extracting the 'maal' slowly kills the soil."

(Tamang 1993:4, 7, 8, 9, 15)

Replenish the plant nutrients removed in harvests, by adding organic materials such as composts, and complementing these, as necessary, with mineral fertilizers (Box 7), and leguminous crops (Plate 36). There should be sufficient nutrients to satisfy not only the needs

PLATE 32
Symbiotic bacteria, chiefly associated with leguminous plants and occurring in root nodules, enrich soils by adding nitrogen, a key plant nutrient. Nodules containing bacteria on the roots of a Vetch plant
[T.F.Shaxson]

of the plants themselves, but also those of the micro-organisms involved in organic-matter transformations and N-fixation (Garrity 1991) (Plate 32).

To restore soil productivity may require regular periods of recuperation during which organic materials are added to increase soil organic matter and biological activity. There are four main ways of increasing the effectiveness of recuperative periods (Shaxson 1993b:111-112; FAO 1995:6):

- increase the duration of recuperative periods;
- reduce the duration of the crops or grazing periods that cause damage to the soil;
- reduce the severity of damage (e.g. by reducing the frequency of tillage or by changing the type of tillage implement).
- manage the vegetation and crops so as to produce moist, adequately-aerated and sufficiently-nutritive soil conditions that will promote biological transformation of organic materials into soil organic matter and humic gums for the improvement of soil architecture.

The recovery of 'soil health' is considered to be very dependent on the presence of active soil organisms, therefore the maintenance of biodiversity among the soil-inhabiting invertebrates and micro-organisms needs special attention (Greenland and Szabolcs 1994:x *et seq.*) (Plate 30).

Consider the growing vegetation itself (crops, pastures or trees) as a major potential contributor of organic materials *in situ*, not only through the production of above-surface litter but also by providing below-surface root residues, which can rapidly restore soil productivity.

Illustrations

Examples of practices that restore soil productivity are illustrated in Plates 33 to 41.

PLATE 33
This farmer's farm had lost all its topsoil in several places due to erosion. In front of him is the area where he has 'constructed' soil for crop production. Initially with crowbar and hammer he breaks up the weathering rock into small pieces about 1 x 1 x 1 cm, then tethers his cow to a stake in the middle of the area and brings it fodder from else-where. The cow adds manure and by walking around crushes the rock. In a year, enough 'proto-soil' has been made for planting the first crop. San Miguel, Ecuador *[T.F. Shaxson]*

PLATE 34
The crumbly absorptive condition of this soil derives much from the incorporation and gradual decomposition of the organic mulch derived from wheat-straw residues in a direct-drilling system. Paraguay *[T.F. Shaxson]*

PLATE 35
Ploughing a sandy clay loam soil for a fourth crop of continuous tobacco, without including a soil-recuperating fallow period under a restorative crop, resulted in many large clods being formed, requiring much additional tillage to achieve a satisfactory tilth. Chiradzulu, Malawi *[T.F. Shaxson]*

PLATE 36
An adjacent plot, the same as that in Plate 35 (note the same shrub in the background) had also just been ploughed for another tobacco crop, but after three years of recuperation under a dense planting of Weeping Love grass. The dense deep rooting of the grass resulted in excellent soil conditions on being ploughed up. The farmer said this plot would need very little further preparation before it was ready for planting. Adequate soil restoration improved the soil as a rooting environment, and resulted in greatly reduced energy requirements and land preparation costs. Chiradzulu, Malawi *[T.F. Shaxson]*

PLATE 37
The exposed white subsoil, shown here, is useless for farming and needs to be recuperated. When this photo was taken, the farmer had already commenced soil recuperation, having constructed contour ditches, planted grass barriers along the contour, applied lower amounts of organic materials, and planted maize. Honduras *[R. Bunch]*

PLATE 38
The same field after heavy applications of organic materials. The field produced about 4.6 t/ha of maize. The beans in the centre produced better than the average yield in the USA. The maize in the upper part is growing on land which had been planted the previous year with Velvet Bean (a high-bulk creeping legume) that was turned under in an 'in-row tillage system'. The lower patch of (sparse) maize did not have the benefit of the Velvet Bean crop in the previous year. No inorganic fertilizer was used on these crops. Honduras, same as Plate 37 *[R. Bunch]*

PLATE 39
The farmer conducted his own experiment to see if the claims of great benefits from organic-matter application were true. He made a number of compost heaps, and buried the material in a strip (along the left-hand side of the photo). Then he planted maize in the traditional manner, one row over the compost, and two rows (to the right of it) on the untreated land. The only difference in treatment between these two rows (the closer one ca. 130 cm high) and the further one (ca. 30 cm high) is the compost and the turning-over of the soil. Honduras *[R. Bunch]*

PLATE 40
This area of communal grazing on shallow stony soil never had a respite from grazing by cattle, sheep, goats. The soil surface was compacted and plant height was kept continually very short and sparse *[T.F. Shaxson]*

PLATE 41
The nearby village community agreed to protect it from uncontrolled grazing while newly-planted young trees were growing. After 18 months the same place looked completely different. The grasses had a chance to seed, and provided more protective cover; the more vigorous plants provided more organic matter both above and below the soil surface; infiltration increased and runoff was reduced. The self-recuperative capacity of the soil:vegetation complex, when adequately husbanded, was startling. Nainod, India *[T.F. Shaxson]*

CONCEPT D

Old approach: Constructing cross-slope physical conservation works to retain soil, water and nutrients is a pre-requisite to obtaining significant yield increases.

New approach: Improving soil management to maximize rainwater absorption and retention will indirectly conserve soil ('by stealth'), and will ensure greater and more immediate yield benefits.

Comments

In dryland farming, soil moisture deficiencies often induce plant-water stresses that are sufficiently severe to affect a plant´s morphological development and which, depending on the growth stage when the stress occurs (see Chapter 3, Fig. 5), may result in yield reductions. The most sensitive growth stages to moisture stress for a number of crops are given by Doorenbos and Kassam (FAO 1979:36).

Although moisture stresses cannot be avoided under rainfed conditions, the frequency and severity of crop moisture deficits can be minimized through good soil management which maintains a protective cover on the soil surface to reduce evaporation losses, and produces optimal soil architectural conditions for the acceptance and retention of rainwater at plant-available tensions.

A lack of rainfall for a sufficiently long period may qualify as an unavoidable climatic drought. But if the soil surface is relatively impermeable, and a significant proportion of the rainfall cannot infiltrate, there will be an additional induced 'subsurface' drought which will worsen the effect of the climatic drought (Plates 42 and 43).

Soil moisture deficiency and any related plant-water stress have relatively immediate effects on plant growth during the growing season, whether soil erosion occurs or not. Effects due to the loss of plant nutrients by erosion, on the other hand, may not be manifest as a nutrient deficiency until the next crop is planted, or even many years later.

PLATE 42
This trash line looks as if it was installed for conserving soil, which it has done to some extent. But there has evidently been another effect of increased rainfall infiltration, increased soil moisture, and hence better growth of the seedlings nearest to the trash line. In this area, farmers move the position of the trash lines from year to year, treating them as mobile compost-strips. Siakago, Kenya *[T.F. Shaxson]*

PLATE 43
In 1994/95 there was almost no rainfall at this site. The runoff plot on the left corresponded to farmer practice of tillage and haphazard planting of maize, without physical banks on the contour or contour-planting. In the middle is a non-experimental area. On the right is a runoff plot with grassed narrow-based banks and contour-planted maize (now-harvested). The photo was taken long after harvest, but the person is indicating the approximate average

heights of maize at harvest. There was practically no runoff from the right-hand plot, but markedly more from the farmers' plot on the left. Because there was so little rain, practically no soil was eroded from either plot. Difference in plant growth between plots could not be equated with differences in soil loss, but was related to the amounts of rainwater retained. Jadan, Ecuador *[T.F. Shaxson]*

PLATE 44
The roots of the wheat plants on the left show the benefits of growing in soil with good architectural conditions, achieved with minimum tillage. Those on the right are from an area of more compacted soil, of poorer architecture, produced by conventional tillage using disc implements. Ponta Grossa, Brazil *[T.F. Shaxson]*

The greater the extent of a plant´s vertical and lateral rooting, the larger will be the store of plant-available water that can be tapped, and the less frequent or severe will be the effects of climatic and induced subsurface droughts (Plate 44).

Even though some annual crops are able to root to 2 metres or more, they are more susceptible than perennial crops to moisture stress, because they only have up to one year for root extension. Some perennial tropical pasture grasses such as *Cenchrus ciliaris* can root to 7 metres or more, and long-lived trees to even greater depths.

In areas where the effective soil depth is less than the potential rooting depth of the annual or perennial plants growing in the soil, climatic droughts are likely to have more severe effects than where soil depth exceeds that of the plants' roots. This limitation is often the case in steepland areas, and any runoff makes the situation worse by diminishing the effectiveness of the rain that does fall.

Improving the infiltration capacity of a soil *in situ,* in areas susceptible to moisture stress, is therefore likely to show more immediate and lasting benefits due to less frequent and less severe moisture stress, than by holding back soil and water at intermittent cross-slope barriers.

The water held up by intermittent cross-slope barriers is water that is already on the move, i.e. it has already been lost from where it originated. Moreover, the water held back at intermittent cross-slope barriers will mostly penetrate the soil vertically, with little lateral subsurface spread, and thus give little benefit to plants in the cropped areas immediately uphill or downhill from the barriers.

The presence of a permeable cover on the soil, such as mulches, is probably more important for its effect of improving and maintaining infiltration capacity and avoiding surface-sealing, than for its simultaneous effect of minimizing the loss of soil nutrients by raindrop-splash.

Subsurface compaction and an increase in pore spaces of very small diameter pores at the expense of larger-diameter pores can also increase subsurface drought by reducing the depth to which plant roots can grow, and possibly also, by increasing soil-moisture tensions to beyond the point at which moisture is readily available to plant roots.

Less runoff (more infiltration) and less erosion are observed from areas with an adequate cover than from areas without any cover which have the 'recommended' frequency of physical conservation banks (De Freitas 1998; Bunch and Lopez 1998; Garrity 1991).

Similarly, soil which moves down-slope and accumulates in narrow strips against the banks of cross-slope physical conservation structures is in the wrong place, and has lost its usefulness, particularly if no attempt is made to return it from where it originated, or to plant crops, grasses or trees on it. Keeping the topsoil in place is not the same as merely 'rescuing' it from muddy runoff water that has accumulated in a hillside ditch. (Note: this is altogether different from the situation where farmers channel runoff and soil sediments to planned areas as in soil-harvesting and water-harvesting.)

Physical soil conservation structures implanted across the slope do not of themselves raise yields immediately up-slope or down-slope of them, apart sometimes from the single rows immediately adjacent to, and on either side of, contour ditches. The better yield of these single rows is attributable to the accumulation of runoff in the contour ditch, and to a lower plant population, since no crop is normally sown in or on the structure itself. The yield increase from these rows is generally offset by the absence of crops where the structure has been installed, and by lower yields further up-slope and down-slope from where water has been lost. By taking strips of land out of the chosen production to construct the conservation structures, and by exposing areas of subsoil materials, the net effects are often to reduce total production (see Box 1).

The process of improving rainwater infiltration through better soil cover and soil architecture in the topsoil not only diminishes runoff and soil loss, but frequently also improves crop growth and yields. Better crop growth will return larger quantities of organic materials to the soil through foliage and root residues, which will enhance soil architectural stability. The associated better root growth of the crop will also contribute to improved soil physical conditions. Thus practices which encourage greater yields also promote better soil architecture, which in turn promotes soil and water conservation. Since the improved conservation is an added benefit to that of increasing yields, the term 'achieving conservation by stealth' is often used.

Principles for fieldwork

Consider the soil as a porous three-dimensional water-harvesting 'sponge', from which water can be recycled to the atmosphere as water vapour through useful crops, pastures and trees, and the balance of which may be collected as liquid water from seepages, springs and deep groundwaters.

Soil qualities which directly influence the availability of plant-available water in soils are rainwater acceptance, storage of plant-available water, adequacy for rooting, and the reduction of evaporative losses. A wide range of soil management interventions, which influence the availability of plant-available water and which are conservation-effective, may be used for different field situations. The relationships between these interventions and soil qualities are shown in matrix form in Table 5 (FAO 1995:26, 27).

Once the soil has been brought into a good porous condition consistent with optimal crop requirements, disturb it as little as possible:

- give preference to minimum- or no-till techniques, over systems that involve frequent tillage which may severely damage soil pore-systems and their continuity with depth;

- protect the surface against the sealing effects of raindrop impact making the best possible use of optimal plant densities, optimum time of planting, leaf-cover, crop residues, tree litter and applied mulches, etc.

Recommend the use of windbreaks where these will minimize the excessively rapid use of soil moisture due to very high evapotranspiration rates, thus postponing the onset of plant-damaging water stresses in the soil.

Use crop varieties and planting dates that will avoid the coincidence of critical growth stages with dekads having a low expectation of rainfall. See Annex 3 on how to calculate rainfall expectations for successive ten-day periods (dekads). For additional information on approaches and techniques for optimal management and use of rainwater, see FAO (1995).

Give more emphasis to demonstrating the advantages of terracing for optimizing the absorption and retention of soil moisture in steep dryland areas, and less emphasis to promoting them for 'soil conservation' purposes; this might encourage more widespread adoption of terraces outside irrigated-rice areas.

Illustrations

Crop residues plus no-tillage (Plate 45) and optimum time of planting (Plate 46) enhance the formation of a good ground cover to promote rainfall infiltration.

Changing the land slope by the construction of bench terraces is an effective way of capturing rainfall *in situ* (Plate 47), and of retaining irrigation water in paddy fields (Plate 48). Where free water is essential for growing rice, farmers are often very skilled in the construction of level bench terraces.

TABLE 5
Relationship between soil management interventions and soil qualities which influence rainwater acceptance and retention (after FAO, 1995)

SOIL QUALITIES AFFECTED BY POSSIBLE INTERVENTIONS AND RETENTION O = possible or most direct effect	Notes	A ACCEPTANCE OF RAINWATER A1 Detention till soak-in A2 Resistance of soil surface to rain impact A3 Rates of infiltration, percolation B STORAGE OF PLANT-AVAILABLE WATER B1 Effective depth of profile B2 Retention per unit depth C MINIMIZE EVAPORATIVE LOSSES C1 From soil surface C2 Through crops, weeds D ADEQUACY FOR ROOTING D1 Aeration D2 Nutrition D3 Root extension									
INTERVENTIONS IN THE FIELD		A1	A2	A3	B1	B2	C1	C2	D1	D2	D3
STRUCTURES											
Level/backslope bench terraces	1	O			O						
Intermittent indiv. terraces, pits + microbanks, etc.	2	O									
Ridge-&-furrow - cross-ties on contour	3	O									
Consn. banks on gradient - waterways, re runoff	4	O			O						
MECHANICAL INTERVENTIONS											
Light scarification of surface	5			O			O				
Rough-surface tillage on contour (plough or tines)	6	O		O		O			O		
Subsoiling to break subsurface compaction	7			O	O				O		O
Change from discs to tines - reduce sub. compaction	8			O		O			O		O
Reduced or minimum tillage with tines, sweeps	9	O		O		O			O		O
Subsurface weeding with horizontal blades	10		O					O			
Alter sand:silt:clay:loam ratios by deep ploughing	11			O		O					O
Minimize unnecessary speed, frequency of tillage	12			O							
Optimize tillage timing versus moisture content	13			O					O		
Import topsoil materials to add soil depth	14				O					O	O
Reduce or minimum surface trampling by livestock, etc.	15			O		O			O		O
PLANTS AND ORGANIC MATERIALS											
Planting along contour	16	O									
Greater/optimum plant density of crops, pastures	17	O	O	O			O				
Crop residues retained - reduced/no-till system	18		O				O	O			
Tree litter retained on surface, not burnt/eaten	19	O	O				O				
Manures, composts, crop residues incorporated	20			O		O			O	O	
Deep/aggressive-rooted crops	21			O	O	O			O		O
INORGANIC MATERIALS											
Chemical fertilizers for macro & micronutrients	22									O	
Liming materials	23			O		O			O		
PROTECTION											
Windbreaks and shelterbelts	24						O	O			
Shade over crops	25		O				O	O			
Pruning	26							O			
Weed control	27							O			
MANAGEMENT											
Regular managed soil-recuperation periods in rotation	28			O	O	O			O	O	
Good husbandry & regular maintenance of improvements	29	O	O	O	O	O	O	O	O	O	O

NOTES ON TYPES OF INTERVENTION SHOWN IN TABLE 5

STRUCTURES

1. Retention of water on surface. Soil depth differs from front to back. Permeability, etc. varies
2. As for 1, but intervening areas preferably covered in dense sward.
3. Surface detention of water. Infiltration better if furrows ripped, subsoiled. Beware localized waterlogging in slowly-permeable soils increases field's surface area for evaporation.
4. By minimizing concentration of any runoff, reduces loss of soil depth by rilling and gullying.

MECHANICAL INTERVENTIONS

5. Breaks soil crusts to increase infiltration. Dust mulch may reduce evaporation from bare soil but is more easily eroded by wind or water.
6. Increases surface roughness and water detention. Increases very large soil voids.
7. Breaks subsurface compacted layers (induced or natural). Increases large soil voids.
8. Tines avoid subsurface compaction caused in certain soils by tillage with discs or hoes.
9. Reduces damage to soil structure. See 6. Best in conjunction with retention of residues on surface.
10. Maintains protective residues on surface: includes no-till/non-inversion/rain-accumulating fallows.
11. Better materials, especially clay, an be brought up to mix in surface layers.
12. Lessens surface pulverization or subsurface compaction and loss of soil voids.
13. Reduces compactive loss of soil voids when soils are tilled or loaded when too wet.
14. Replacement of eroded materials from downslope, or enrichment with alluvial materials, etc.
15. Once infiltration rate restored, limits compactive loss of soil voids, maintains infiltration through surface.

PLANTS AND ORGANIC MATERIALS

16. Favours inter-row tillage on contour, cross-slope roughness. Residues may lodge against stems as micro-carriers to runoff.
17. Improves barrier effect (16) to cover against raindrop impact. Associated organic residues and roots favour percolation, structure improvement, aeration, etc.
18. Improved protective cover to soil, added organic materials, soil voids maintained. Improves catchment hydrology.
19. As 18. Main benefit of trees in hydrologic cycle.
20. Multiple effects: substrate for soil organisms, eventually affecting soil structure, aeration, water absorption. Ease of seedbed preparation. Slow-release plant nutrition. Buffers effects of inorganic fertilizer and increases efficiency of use.
21. Compacted, dense layers are penetrated, subsequently leaving channels for air, water, root extension.

INORGANIC MATERIALS

22. Satisfy absolute deficiencies, optimize nutrient balances, favour root extension and function.
23. Flocculation of clays gives structure improvement: lower soil acidity improves nutrient ion availabilities: more micro-organic activity: Ca supply.

PROTECTION

24. Reduce direct evaporation and transpiration losses, prolonging periods before onset of water stress in plants. (Beware root competition for moisture and lateral spacing.)
25. As 24.. Also limits leaf temperatures: may provide organic litter. Both may compete for soil moisture.
26. Where feasible, reduces transpiring leaf area, slowing onset of water stress.
27. Cuts moisture loss through unwanted plants. Where feasible, retain on surface as dead mulch.

MANAGEMENT

28. Grasses and legumes managed for soil structure renewal provide fibrous roots. N-fixation, improved macro-, micro-organic activity, renewal of soil voids, better water retention and bring up nutrient materials from deep soil layers.

PLATE 45
A well-grown crop of maize direct-planted through a mulch of residues from the previous crop. Good cover is provided by the combination of crop leaves and surface residues. Santa Catarina, Brazil *[T.F. Shaxson]*

PLATE 46
Two adjacent fields of sorghum, belonging to different farmers. Differences in planting dates and crop management by the two farmers have resulted in marked differences in crop growth and soil cover, and hence in the severity of damage to the soil by high-energy raindrops. The same considerations apply on steeplands. Indore, India *[T.F. Shaxson]*

Where bench terraces are not used, the effects of an abundance of crop residues left on the surface of a permeable soil may be more effective in capturing rainwater and increasing soil moisture than the construction even of large transverse conservation banks (Plate 49).

PLATE 47
In some places slope steepness is lessened for non-irrigated crop production by constructing bench-type terraces which may follow the contour less strictly. Again, their purpose may be more oriented to capturing rainfall to increase soil moisture than to 'doing soil conservation' *per se*. Ashapuri, India *[T.F. Shaxson]*

PLATE 48
Level bench terraces are constructed and maintained where water control is paramount - the farmers almost certainly do not construct them for 'soil conservation' purposes.Nepal *[Natural Resources Institute]*

PLATE 49
Physical conservation works (here a large 'murundum') were formerly considered the key parts of a soil and water conservation strategy. Both research and field experience have shown that attention to breaking the force of raindrops falling in the vertical dimension, using crop residues in particular, is more effective in achieving conservation of water (and consequently of soil) than relying only on contoured physical works which are only effective against runoff in the lateral dimension. The crop residues on the right of the picture may be more cost-effective than the large earthwork on the left, as long as the soil beneath the residues is capable of absorbing all the rainfall. Cascavel, Brazil *[T.F. Shaxson]*

Figure 9 shows how leaving crop residues on the surface (as in no-till) without any terraces can result in much less runoff than terracing in the near absence of crop residues (as in ploughing).

CONCEPT E

Old concept: **Sustainability of land productivity is achieved by the installation of intermittent cross-slope physical structures or vegetative barriers, combined with adequate fertilization.**

New concept: **Sustainability of land productivity is achieved by continuous active conservation-effective land management (land husbandry) over time.**

Comments

The one-off implementation of soil and water conservation measures combined with adequate fertilization is not sufficient to sustain land productivity. All the physical, architectural, chemical and bio-logical attributes of soil, which contribute to its productivity, sustainability and usefulness, must be considered in a holistic manner.

Physical conservation structures can only function as they are designed to if the management of the land above them is of an excellent standard. If it is not, the physical structures are likely to fail more frequently than intended by their design life, through silting-up, over-topping and/or collapse. Installing more of them is no substitute for better land management over the whole of the field.

The use of land for sustained agricultural use requires more than a particular type of management aimed at producing a specific output.

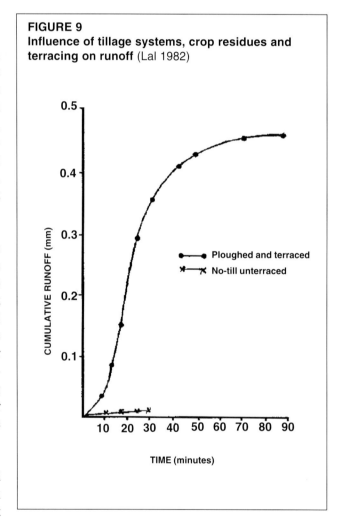

FIGURE 9
Influence of tillage systems, crop residues and terracing on runoff (Lal 1982)

Farmers are constantly being confronted with changes in many variables: markets, production technologies, prices, the availability of supplies, access to transportation, credit availability, income from non-farm activities, labour supply, the climate, and other environmental, social, and economic factors. It is therefore unlikely that a farmer will be able to continue a specific system of production and land management for a prolonged period. Changes will be inevitable, and as a result land-use and management practices are also likely to change.

Sustained productivity from land requires good, active, management, and a concern not only for the quantity, quality and stability of the outputs, but a concern also for the land resource, i.e. the quality of the land and the health of the soil. To embrace these ideas the term 'land husbandry' is increasingly being used. Land husbandry is not confined to the notion of management, it also includes the notions of 'concern for the land' and 'taking care of the land' in much the same way that animals or crops are treated with concern and cared for in animal or crop husbandry. A suitable definition of land husbandry (which is a modified version of that given by Downes 1982:126) is:

Good land husbandry is the active process of selecting and implementing systems of land use and management in such ways that there will be an increase in, or at least no loss of, land quality and soil health as well as land productivity, stability and usefulness.

The term 'husbandry' also connotes the idea that the farmer's livelihood is entwined with the health of the land, rather than being separate from, and imposed upon, the land, and alludes to the emotional linkages that exist between people and the land they occupy and depend upon.

Women and men in farming communities may conceive land husbandry differently, and their different perceptions may give rise to contrasting opinions regarding how their land resources should best be used.

Principles for fieldwork

Help small-scale, resource-poor farmers to improve their skills in land husbandry, and thus the sustainability of their chosen enterprises. This will generally prove much more acceptable, and will also give better value for money and effort, than merely implementing physical conservation structures, whose benefits cannot be clearly or rapidly perceived, or demonstrated.

Farmers are unlikely to make significant changes to what they are already doing merely for 'soil conservation'. But improvements in management, which help them achieve their aims of more reliable yields and greater production per unit area, may, if chosen appropriately, increase the conservation-effectiveness of their existing farming or land-use systems.

'Improvement' may be only the betterment of present methods of management, but it may also include beneficial add-on practices which the farmer is not yet using. A hierarchy of possible improved and add-on practices that are conservation-effective is presented in Table 6. They have been ranked from those requiring the least drastic, to those requiring the most drastic, changes relative to existing farming practices.

Assist farmers to appreciate how soil-related constraints to productivity may be more effectively overcome by the judicious selection of particular combinations of land management practices. For example, the combination of optimum time of planting and adequate fertilization can greatly affect the development of crop cover, as illustrated in Figure 10, which in turn will affect rainfall infiltration and soil conservation.

Suggest any necessary physical conservation structures, not as independent add-on features, but as complementary practices to good land husbandry. These may be needed to mitigate the effects of infrequent but particularly severe rainfall events, or to stabilize land that is in process of being reclaimed, but which is becoming degraded more rapidly than vegetation can cover and protect it. Physical conservation structures must never be regarded as substitutes for good land husbandry.

TABLE 6
Hierarchy of conservation-effective actions (Set A = least drastic; Set D = most drastic)

SET A: IMPROVEMENTS IN CROP-, PASTURE- AND TREE-MANAGEMENT PRACTICES WHICH ARE PART OF THE CURRENT FARMING SYSTEM, AND ARE CONSERVATION-EFFECTIVE BECAUSE OF:

I. SOIL-COVER EFFECTS (reduced rainfall erosivity, surface windspeed)

Improve present practices	Add-on practices
- Crop/pasture/tree cultivars - Intercropping - Under sowing - Stubble, residues, no-till - Grazing management - Rotations - Optimize plant population/ha	- New species - Fencing for pasture subdivision - Shelter belts, windbreaks - Cover or catch crops - Planted fodders - Applied mulch

---

II. SOIL QUALITY EFFECTS (improved physical, chemical and biological characteristics of soil for better rooting, greater infiltration and lower erodibility)

Improve present practices	Add-on practices
- Reduced/minimum/zero tillage - Tillage with tines - Lime, fertilizers - Manures, composts - Lessen causes of soil compaction - N-fixing crops, green mulches (minimum tillage) - Rotations - Rough-surface tillage practices (chisel tines) - Planting on contour - Improved fallows	- Subsoiling, 'pan'-breaking - Periods x management for soil recuperation - Cover/green-manure crops - Add woody species (agro-forestry) - Better control of fire

---

III. TIMELINESS EFFECTS (improved timing of operations, speed, type of action, ease of management)

Improve present practices	Add-on practices
- Optimum planting dates - Optimum dates to control pests, weeds - More suitable breeds of animals - Feeding - Yokes, harness - Equipment for tillage/sowing/weeding/harvest transport/storage/processing - Road maintenance re market access - Long/short-season cultivars	- Tractor and equipment - Road realignment - Planted fodders, browse, hay silage

---

SET B: BETTER MATCHING OF LAND USE TO AGRO-ECOLOGICAL NICHES IN LANDSCAPE ON FARM

Improve present practices	Add-on practices
- Land-use planning (spatial) - Land-use reallocations to optimum places - Optimum balance of current types of use	- Better management of unused land - Alter types of output - More diverse crops

TABLE 6 Cont'd

SET C: ADDITIONAL CONSERVATION-SPECIFIC PRACTICES	
	Add-on practices **To catch/slow/divert runoff** - Dams/tanks/ponds - Bench terraces - Contour absorption ditches, graded hillside ditches - Water harvesting/spreading techniques - Storm drains, cutoffs, diversions - Conservation banks on gradient - Lock-and-spill drains (acequias + gavetas) - Orchard terraces - Controlled-gradient roads - Filter strips, live barriers - Ridging, tied-ridging - Pasture furrows - One-way up-slope contour ploughing - Use of reversible plough **To trap soil** - Conservation banks on contour - Silt-, sediment-traps - Filter strips, live barriers - Stone walls **To lessen slope of area to be used** - Land shaping, levelling - Bench terraces backward/forward sloping - Weirs, checkdams to lessen channel gradient **To stabilize drainage base level** - Weirs, checkdams - Waterways of grass, stone **To prevent mass-slippage** - Streambank protection - Cutoffs, diversions - Dewatering - Deep-rooted plants - Stabilization of seepage through dam walls **To prevent flooding** - Drains, levees
SET D: RECLAMATION OF BADLY-DEGRADED AREAS (BADLY-ERODED, WATERLOGGED, SALINIZED)	
	Add-on, changed practices - Use appropriate combinations of above practices - Change to protective/recuperative types of land use and management - Vegetative/physical stabilization of gullies - Reduce excessive irrigatio - Improve land drainage - Re-introduce varied plant species - Cut-off diversions - Fencing to exclude grazing animals

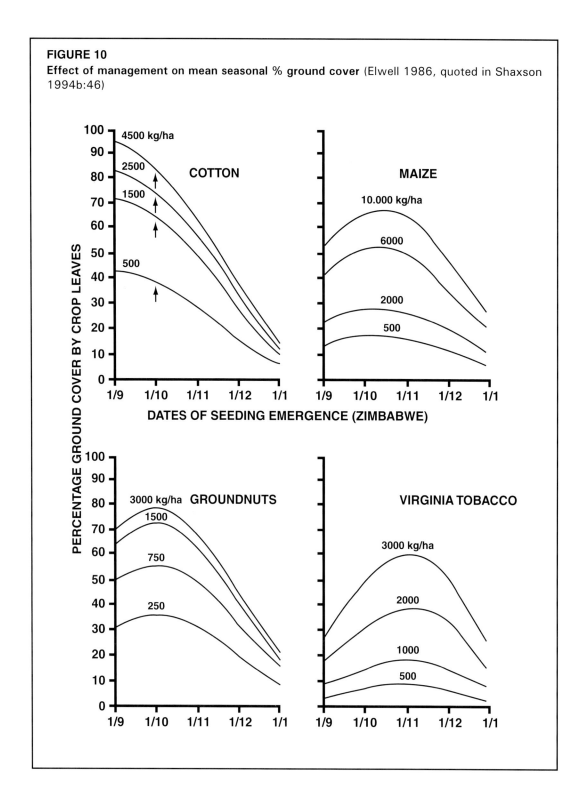

FIGURE 10
Effect of management on mean seasonal % ground cover (Elwell 1986, quoted in Shaxson 1994b:46)

Consider how regular farming activities might be improved by better husbandry to achieve greater environmental, social and economic benefits, e.g. tillage; control of pests, diseases and weeds; pasture management; fodder production; use of residues; complementing plant nutrients; managing rainfall and rainwater; regular soil self-recuperation; producing more abundant crop growth.

Expose farmers to a variety of farming enterprises and farming situations through visits and tours to increase their awareness of the need for land management practices to be modified and changed according to the production system, environmental, social and economic factors. In this way farmers will become better prepared psychologically to adapt their land-use and management systems in response to changing circumstances.

Illustrations

Plate 50 shows the benefits of complementing good cover management with physical structures to mitigate severe rainfall events in the commercial production of tea.

PLATE 50
A combined system of in-row mulching and inter-row tied-ridging for contour-planted tea ensures maximum water absorption (and consequently maximum soil conservation). Continuous mulch-cover would present a serious fire hazard to young tea plants. Lujeri, Malawi *[T.F. Shaxson]*

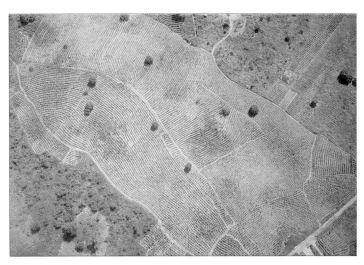

PLATE 51
Aerial view of a poor stand of village maize - probably due to a combination of insufficiency of nutrients, delayed planting after the optimum date, poorly-adapted variety and other problems within the farming system Toleza, Malawi *[T.F. Shaxson]*

The need for the judicious combination of management practices to achieve good land husbandry is shown in Plates 51 and 52, and examples of the need to adapt management practices, in this case tillage, according to environmental conditions are presented in Plates 53 to 59.

PLATE 52
From the same altitude as Plate 51, a few kilometres further on, a commercial planting of maize with adequate timing, inputs, and management. Note the marked difference not only in likely output but also in the capacities of the maize (same variety, different management) to provide protective cover against rainfall impact. Toleza, Malawi, 1971 *[T.F. Shaxson]*

PLATE 53
A combination of practices which together favour the rapid growth of young tea plants, and at the same time secure the land against extreme rainfall events. On this tea estate it was possible to design the actions from the beginning: natural drainage lines and crests were identified and reserved for waterways and road lines respectively; conservation banks on a slight gradient link with the roadside drainage and carry unavoidable runoff across the slope. But the key components are the combination of mulch and micro-catchments (tied ridging), as in Plate 50, within which the young bushes were planted, and from which very little runoff originated. The system was adopted because it simultaneously maximized capture and infiltration of seasonal rainfall to sustain the plants over the dry season, and minimized problems of keeping earth roads passable in all weathers. Protection against the drying effects of high winds is provided by the straight lines of shelter-belts which cross the area. The estate management wished to reduce the unit costs of tea production by raising yields and reducing running costs. The manner in which it was carried out also allowed conservation to be achieved 'by stealth'. Lujeri, Malawi *[T.R. Jackson]*

PLATE 54
This farmer decided it was safer to plough up-and-down slope than to drive along the contour with the risk of over-turning the tractor. However, the resulting furrows favour the rapid flow of runoff water and loss of soil. Using a tractor on such a steep slope is NOT compatible with good husbandry. Burgay, Ecuador *[T.F. Shaxson]*

PLATE 55
As a result of ploughing up-and-down the steep slope, runoff water has concentrated in the furrows and eroded the soil. Burgay, Ecuador *[T.F. Shaxson]*

PLATE 56
Animal traction is an alternative option for contour-ploughing, especially on steeper slopes. Caxambu, Brazil *[T.F. Shaxson]*

PLATE 57
A management decision about types of tillage affects the soil architecture and structural condition. This cloddy condition is the result of long-continued tillage with tractor and heavy disc implements, and without retention of crop residue mulch. Carambei, Brazil *[T.F. Shaxson]*

PLATE 58
In another area on the same farm, a decision to adopt a no-till system resulted in significant improvements to the soil as a rooting environment, and in its capacity to absorb rainfall and thus minimize runoff. Carambei, Brazil *[T.F. Shaxson]*

PLATE 59
Using a creeping vetch as a living mulch through which maize was planted resulted in marked changes in soil colour, stability of soil architecture, N-content, maize yield and year-round protection of the soil surface (see also Plate 23). Chapeco, Brazil *[T.F. Shaxson]*

PLATE 60
This farm-family decided to make best possible use of manures and composts, using all available organic matter, to improve the soil for crop production. The results of this intensification mean that total farm production has risen. By being able to produce a constant amount of, say, maize, from a smaller area than before, it has become possible to diversify into other crops. Mukhumu, Kenya *[T.F. Shaxson]*

Optimizing the use of crop residues and the introduction of cover crops are key components in good land husbandry, and can often result in more intensive production of traditional crops, so that more land becomes available for the introduction of new 'diversifying' options (Plate 60).

Concept F

Old approach: **Devise land management practices for each enterprise separately.**

New approach: **Devise land management practices that consider the whole farming system, so that implications to land management from interactions between components of the farming system are fully taken into account.**

Comments

Farming systems are generally more complex, and often consist of a larger number of component activities, than is generally appreciated by outsiders.

Many of the component activities within a farming system may be interrelated due to the constraints of land area, location, water supplies, cash-flow, labour, or equipment. For some component activities of the farming system, the outputs of one activity may be utilized as inputs to another activity.

Because of the interrelated nature of the components of a farming system, changing the management of one component may directly or indirectly influence the feasibility, productivity or profitability of the other components.

Therefore the change in concept from considering land management for a particular crop to land management for the whole farming system, usually requires a change from a mono-disciplinary to a multi-disciplinary approach.

The nature of western scientific research and training is towards specialization in particular academic 'disciplines', and the imparting of specialized and disciplinary knowledge to field staff in agriculture. Often this is not appropriate to the complex and interrelated nature of the assorted problems that confront farmers who deal with farming systems. Extensionists should therefore have a multidisciplinary training in order to address satisfactorily the problems of farming systems.

Principles for fieldwork

Provide farmers with a better understanding of the interactions between the various components of the farming system, the possible problems and potentials, so that they can make improved decisions on how best to manage the interrelating components – plants, animals and land.

Help farmers to recognize and understand any conflicts that may arise in the management and use of resources within their farming system which could lead to environmental deterioration, and the options that may be used to overcome these conflicts.

Encourage farmers to appreciate how to use more efficiently, and in a more integrated manner, the outputs and 'waste' products from one component of the farming system for the benefit of other components.

Illustrations

The excessive grazing of crop residues by livestock in Kenya has led to the development of practically denuded grazing lands in Kenya, which are highly susceptible to soil erosion (Plate 61). This problem can be best solved by considering the whole farming system. By introducing some protection for the natural vegetation to allow for natural regeneration, and achieving a controlled degree of grazing, a balanced situation has been reached between production and protection, so that the pastures provide both fodder and ground protection (Plate 62). The important role played by fencing in controlling grazing intensity so that an adequate soil cover is maintained for conservation purposes is shown in Plate 63.

In many countries a serious problem exists in the fate or utilization of crop residues, which can best be solved using a farming systems approach. Crop residues are needed both for soil protection and also frequently for dry-season fodder.

PLATE 61
A bare land surface where overgrazing has removed much of the soil's protective cover, though protection from a high stone cover still remains *[T.F. Shaxson]*

PLATE 62
Where some protection from grazing has allowed the vegetation to recover, the grasses once again are sufficiently productive to provide both fodder and ground cover. The balance achieved between production and protection depends on the skill of managing the dynamic system of rainwater - soil - vegetation - animals, Tharaka-Nithi, Kenya *[T.F. Shaxson]*

PLATE 63
A fence is a tool in husbandry, with effects not only on grazing intensity but, indirectly, on the conservation of water and soil. This fence separates an area under conservative grazing management (left) from an area that has been more drastically grazed. Molimo Nthusi, Lesotho *[T.F. Shaxson]*

Studies in El Salvador have shown that a 75% ground cover effectively reduces erosion hazards to low levels, and by encouraging the production of additional fodder from silage, forage trees in live fences, and improved pastures, more of the crop residues can be left on the soil surface to enhance rainfall infiltration, and reduce runoff and soil losses (Barber, 1998).

The relationships between the actual production levels of residues, the quantities required to satisfy the requirements for soil protection and livestock fodder, and the extra quantities of dry matter that need to be produced from additional sources are shown in Figure 11.

Adjusting one component of a farming system can bring about several benefits, as in the case in Jamburdi Hapsi, India (Plate 64), where replacing a long-duration sorghum variety with a short-duration variety permitted the introduction of a second crop, which more than doubled total yield, gave more cover, and increased soil protection.

PLATE 64
On the left is a long-duration sorghum traditional variety not yet ready for harvest. On these vertisolic soils it only matures after the end of the rainy season and, by transpiring residual moisture, precludes the chance of planting a crop of e.g. wheat in the dry season. On the right are bundles of an improved composite variety which matures quicker, does not use up much residual soil moisture before harvest, and thus leaves enough moisture for a follow-on crop in the dry season. Whereas formerly, following the traditional variety, farmers would keep the land weed-free throughout the dry season and the following wet season, to replenish water supplies, with the new short-season variety much more efficient use of the land is possible. Output per ha more than doubled, and soil protection increased because in the rainy seasons the land always has a crop cover every year. A change in the variety of sorghum had multiple benefits on the farming system. Jamburdi Hapsi, India *[T.F. Shaxson]*

FIGURE 11
Relationships between the factors influencing residue production and residue utilization
(Barber, 1998)

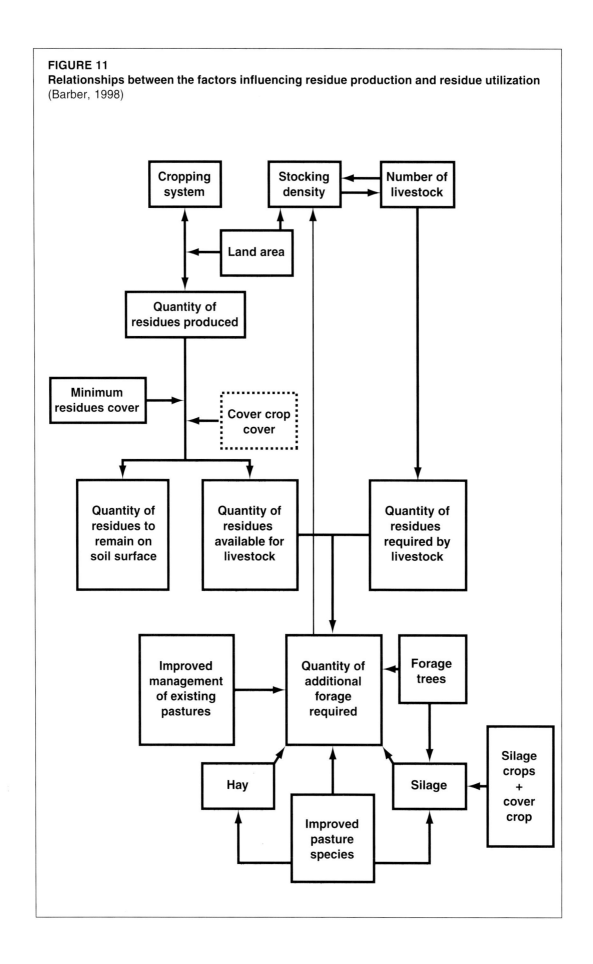

PLATE 65
The farmer experimented using 'plant tea' - the liquid from macerated leguminous plants stirred with water and left to ferment. When the maize was still at seedling stage, a few cm high, the maize on the right received one cupful of the liquor per plant; that on the left (planted at the same time) did not. Management decisions that place greater reliance on the recycling of organic materials and stimulating biological activity in the soil provide significant opportunities for improving the soil and its productivity, and for increasing the conservation-effectiveness of farming operations. Kakamega, Kenya *[T.F.Shaxson]*

The recycling of 'waste' organic products produced within the farming system may also bring about substantial and multiple benefits as in the case of 'plant tea' used in Kenya (Plate 65).

CONCEPT G

Old concept: **Resource-poor, small-scale farmers are naturally conservative, irrational, ignorant and unconcerned about the productivity of their land, and thus are part of the problem of land degradation.**

New concept: **Farmers make rational decisions within the envelope of constraints in which they live, and are vitally concerned about the productivity of their land. They should therefore be considered as part of the solution to the problem of land degradation.**

Comments

Resource-poor farmers are more vitally concerned about maintaining the productivity of their land than any outsider. If farmers´ land is now degrading, the problem is not that they do not care (as has so often been assumed), but that they cannot find appropriate solutions among the recommendations and suggestions available to them. It may be that suitable recommendations do not exist, but even if they do, farmers may not be aware of them because they seldom if ever meet an advisory worker.

Farming families have detailed local knowledge of their surroundings, both social and environmental, and understand how these aspects are likely to react to changes. This is likely to temper their initial enthusiasm for recommendations, the effects of which they have not yet been able to test or judge from friends' or neighbours' experiences.

The rejection by farmers of what appear to be logical recommendations may give the impression that they are innately conservative, reactionary, or stupid, whereas their rejection may be the understandable result of an astute comparison between the likely risks, costs or sacrifices of introducing a recommendation, and the possible benefits. For small-scale, resource-poor farmers the price of failing to produce sufficient food is destitution!

Farmers´ views on what contributes to, and is evidence of, loss of productivity often appear to differ from the views of outsiders. They are usually worried if their yields appear to be falling year by year. However, in any year they are generally much more concerned about the amounts of seed, manure and fertilizer that have been washed away, than they are about the loss of soil particles *per se*, even if soil erosion and runoff are evident (White and Jickling 1994:99+).

The interests of male farmers often focus on the maintenance of yields and the commercialization of produce. In rural communities the women may have different knowledge, experience, and opinions; they are often more concerned with the health of the land, and the quality of its products in relation to the 'life-forces' of family members and animals in their care. In village communities women often prove to be better 'vigilantes' of natural resources than men.

Rural people in the tropics and subtropics have commonly been perceived as the causes of land degradation and condemned accordingly. This attitude towards rural people, especially in steeplands, should be changed from: 'Land and the people who ruin it' to 'People and the land they occupy, and who would look after it better if they knew how'. "We have to stop thinking of farmers as part of the problem, and make them part of the solution" (Hudson 1988:4).

Principles for fieldwork

In making suggestions for improvements, build on what farmers know and do already, and with which they are familiar, rather than attempt to introduce enterprises or techniques about which they have little knowledge, and for which they cannot realistically assess the risks.

In suggesting an improvement that is known to be conservation-effective, stress more its demonstrative capacity for raising and maintaining productivity than its capacity to reduce land degradation. Technical staff should not 'hi-jack' what are basically good husbandry practices and call them 'soil conservation' practices, because this may make them less likely, rather than more likely, to be adopted.

In making suggestions for improvements, rely as much as possible on the resources readily available to farmers, and consider scarce resources as complementary. This particularly applies to the use of organic materials, both plants and their residues, for restoring soil productivity.

For a recommendation to be successfully adopted, it should be:

• acceptable;
• feasible,
• appropriate;
• beneficial;
• economic;
• effective for infiltration and conservation.

Techniques which prove to be favoured by farmers and which may be widely adopted without external incentives (White and Jickling 1994:102) are those which:

- combine components which are familiar to farmers;
- are compatible with other agricultural and social activities;
- are simple, of low cost, and do not require finance for implementation;
- provide short-term economic returns;
- do not require additional labour;
- are adaptable to specific site conditions;
- can be adapted and adopted sequentially as the farmers' experience of the technology increases.

There are several people-based resources which remain effectively untapped in rural areas, and which should be used to advantage to further the development and uptake of good land management techniques:

- farmers' local knowledge (Plates 66 and 67);
- farmers' inventiveness and capacity for innovation;
- farm families' latent skills and enthusiasms when engaged in activities that concern and interest them;
- the views, concerns and initiatives of women.

PLATE 66
Local people have intimate knowledge of their surroundings, and readily become involved, one with another, in visualizing what they know in the form of maps, diagrams and other forms of representation. This constitutes an valuable activity when carrying out participatory diagnostic studies in a community. Malawi *[J. Pretty]*

Illustrations

Farmers´ practices that seem irrational to outsiders and non-farming agriculturalists may have a sound rationale according to farmers (Box 8 and Plate 68).

BOX 8: WHY MANY TERRACES ARE OUTWARD-SLOPING IN NEPAL

"...they must be aware of the erosion of topsoil from these sloping terraces so why do they allow it to happen? When this was put to (the farmers) they replied: 'Yes, we know that the topsoil in which we have put compost and other ingredients and which is the most fertile is being eroded every year. Yes, we know that because of soil erosion the yield from this terrace is poor. Despite these disadvantages we allow these terraces to be steeply sloped. If the slope were changed and they became level we would gain by reducing soil erosion and better crop yield but we would lose the whole terrace. In June when the monsoon begins all the soils are dry, they absorb much of the early rains becoming wet and very heavy. But the rain continues and in Bhadra (Aug/Sept) the terrace, completely saturated with water, would not be able to withstand the increasing burden and would collapse'."   (Tamang, 1993:12)

PLATE 67
Both women and men have particular knowledge of the agricultural year as it affects their daily lives - here a calendar of events is being produced. India [*J. Pretty*]

PLATE 68
In such steep country 'conventional wisdom' suggests that outward-sloping bench-type terraces would not be useful. But the farmers here have their own explanation of why they make them in this way, and not level or backward-sloping. Nepal [*J. Clark*]. **See Box 8.**

An example of how management can be successfully improved, starting with the farmer's existing technology and making slight modifications, is shown in Plates 69-72. A slight modification of the seeder permitted the application of fertilizer and seed to different depths in the same operation.

PLATE 69
The traditional seeder is made of wood, and has one hopper for seed, (fertilizer, when used, being mixed with the seed) and split between two outlets to the seeding tines. There is often no advantage to applying the fertilizer mixed with the seed in the same row, as this frequently results in poor seed germination *[T.F. Shaxson]*

PLATE 70
It was demonstrated to the farmers that better germination occurred when the fertilizer was banded 1 cm or so below the line of seeds, and that this could be done with a commercially-manufactured combine seeder. The village blacksmith, however, said he could modify the existing seeders to do the same job without much cost or difficulty. The fertilizer on its own is placed down the forward spout and falls slightly in front of and below the seed, which falls down the rear spout *[T.F. Shaxson]*

PLATE 71
At seeding, the system needs one person to guide the bullocks and one to place the materials in the hopper *[T.F. Shaxson]*

PLATE 72
The seeding operation now requires three people, but the better germination and seedling growth obtained from separating the fertilizer band from the band of seed was considered to more than compensate for the extra work. **Jamburdi Hapsi, India** *[T.F. Shaxson]*

CONCEPT **H**

Old approach: **The role of extensionists is that of instructors and advisers who transfer prescribed soil and water conservation technologies to increase productivity.**

New approach: **The role of extensionists is that of promoters and facilitators who assist farmers to identify and solve their own problems to increase productivity.**

Comments

The old concept of extensionists is that they are advisers and instructors who impart knowledge and demonstrate prescribed technologies to farmers in order to overcome land degradation problems and increase productivity.

However, the contexts (social, economic, educational and environmental) within which resource-poor small-scale farmers achieve their livelihoods are frequently complicated, and often markedly different from those of their extensionists.

In order to spread perceived risks due to weather and markets, small farmers' systems of livelihood often have many diversified components, both on-farm and off-farm, which interlink in various ways. Family income may originate not only from the farm, but also from remittances made by family members working on non-farm jobs at home or elsewhere.

Farmers make rational decisions about how to use the resources available to them to achieve their many objectives, but these decisions are frequently affected by various constraints and limitations which may not be fully appreciated or even known by the extensionists.

The constraints and limitations surrounding a farmer are sometimes referred to as an 'envelope' of limitations and constraints. The challenge of an extensionist is not so much to try and change the farmers' rationality, as to try and understand the farmers' rationality, i.e. their perceptions of their limitations and opportunities. In this way the extensionist can more effectively help the farmers overcome as many of their constraints as possible, thus... 'enlarging the size and shape of the envelope', or...giving the farmers greater freedom and flexibility in which to make their day-to-day decisions (Figure 12).

Farming families generally possess the knowledge of at least their forebears' 'traditional' practices which have enabled the family to survive up to the present in both good and bad years. The accumulated knowledge and experience possessed by elders in rural communities are resources not quickly acquired by younger non-farming agriculturalists. Thus new ideas recommended by outsiders will be scrutinized very carefully to ensure that their adoption will not threaten the security of existing systems of production.

With some justification, experienced farmers often view technical staff who are younger than themselves, with different training, and little comparable experience of surviving in difficult conditions, with a degree of polite scepticism, if not suspicion (Herweg 1993: 393).

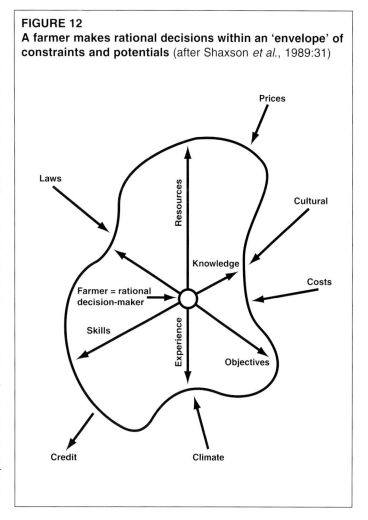

FIGURE 12
A farmer makes rational decisions within an 'envelope' of constraints and potentials (after Shaxson *et al.*, 1989:31)

Extensionists must therefore prove themselves to the satisfaction of farmers before they can gain credibility. They must not assume that their scientific knowledge will automatically give them this credibility.

By adopting the role of promoter and facilitator, the extensionists can more easily develop a two-way dialogue with farmers that will enable them to appreciate and understand the farmers' problems, and how they are perceived by farmers. A two-way dialogue will be more feasible once the extensionist has developed credibility in the eyes of the farmers, and when feelings of mutual trust and of partnership towards the improvement of agriculture have developed.

It is frequently difficult for extensionists to assume the new role of promoter and facilitator when they have been trained as instructors and advisers in the traditional system. The change signifies not only a change in the way they relate to farmers, but also signifies a relinquishment of authority and power in their new relationship with farmers. The extensionists must learn to accept that their perception of a problem or solution may not be the correct one for, or be acceptable to, the farmer.

The extensionist's new role of promoter and facilitator also signifies a change in the concept of farmer participation from 'the people participating in the implementation of plans considered to be good for the farmers, which have been devised by outsiders', to 'extensionists participating with farming families in helping them to identify and rank their most important problems, to decide what activities are needed to overcome the problems, and to implement the activities'.

Participatory activities as used in 'participatory learning and action' (PLA), (formerly known as "participatory rural appraisal") permit a better understanding of people's perceptions and viewpoints, the strategies and techniques they use to cope with different situations, and the beneficial changes they have already accomplished. They also facilitate the identification of farmers´ resources, problems, constraints and potentials, the options for overcoming their problems, and possible opportunities.

Principles for fieldwork

Establish a good two-way dialogue between extensionist and farm-family members that allows for information flow in both directions. This will enable the extensionist to help farmers identify their most urgent problems, and to help them formulate acceptable solutions.

Organize regular discussion sessions in which the problems, possibilities, experiences, hopes and practicalities can be expressed and considered by the whole group of farmers (Plate 73).

PLATE 73
Information can be easily transferred in both directions between farmers and extensionists, once the extensionists have been able to develop credibility in the eyes of the farmers, and when feelings of mutual trust and partnership in the improvement of agriculture have developed. Rinjlai, India
[T.F. Shaxson]

Promote PLA activities. For details of these activities see, for example, IIED 1994, and Scoones *et al.* 1994. The PLA activities will allow insights into such matters as:

* the nature, dynamics and rationale behind local land-use systems;
* farmers' knowledge about natural resources, systems of classification and ranking;
* the nature, causes and rates of land degradation or improvement;
* the distribution of resources and land use indicated by people's mapping of their area (perhaps augmented by shared aerial photo interpretation (Carson 1985);
* the nature and causes of agricultural problems;
* possible solutions to the problems;
* the access to, and usefulness of, advisory services etc.

If data which farmers have provided in participatory diagnostic studies are taken away, collated and analysed by an extensionist or outsiders, the results must always be returned to the

PLATE 74
Project staff have brought back analyses of the farmers' micro-economic data, provoking much interested discussion and growing understanding among the farmers of how such information can help their own decision-making. Labrador, Costa Rica *[T.F. Shaxson]*

farmers for their own consideration and discussion, so that they can take these data into account when planning the activities required to solve their problems (Plate 74).

Learn what criteria farmers use for making their own land classifications. Long-established farmers have an intimate and detailed knowledge of their soils, and classify them in pragmatic terms. While these may be difficult to correlate with those of neighbouring farmers or with pedological or land capability classes, they nevertheless represent a much underrated resource (Pawluk *et al.*, 1992; Whitehouse 1990:54; Douglas 1994b:102,103; Fujisaka and Garrity 1991:223). In particular, these classifications will be based on pragmatic criteria related to crop growth, and the behaviour of soils in different locations (e.g. aspect), and different management systems.

Define and discuss problems together with farmers in the participatory diagnostic and planning events, not independently of them. They will probably identify those which are most immediate and evident; however they may not be able to foresee some potential problems that could arise in the future, or identify certain physical or chemical problems which are not readily apparent. This is a specific role for advisory workers to take up (Vieira and Shaxson 1995:7+).

Identify the real roots of the problems and the contexts in which they occur: they may be very different from what may be supposed by either farmers on the one hand, or field staff on the other (Box 9).

Encourage technical staff from different relevant disciplines to participate in identifying problems with farmers and formulating inter-disciplinary solutions which adequately address the complex interrelationships that exist between the various components of the farmers' livelihood system. See Concept F on farming systems.

Blend together outside technical knowledge and farmers' own pragmatic knowledge, so as to be able to sift and identify all available options for solving the identified problems.

Introduce suggestions of improvements into discussions with farmers as talking points, rather than insisting that certain actions must be taken. Before making suggestions, be sure they can give significant improvements that can be recognized by farmers. Small-scale resource-poor

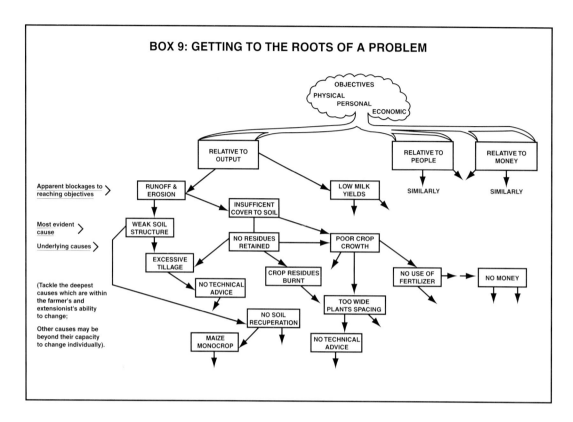

farmers seem unlikely to adopt a new technique unless the net marginal rate of return is greater than about 50% (Hudson 1991:13; Piha 1993:411; Bunch and Lopez 1998). If the suggestions fail, credibility may diminish and will have to be re-built.

Start with the knowledge and techniques that farmers already possess, which can be used as a basis for expanding and enriching their knowledge and skills, rather than bringing in completely new ideas based on untested assumptions about the contexts in which they are to be applied (Premkumar 1994). An example of a farmer's knowledge of the utility of various indigenous plants in a 1-ha area is given in Box 10. Such information could be very useful for the development of improved agroforestry in particular farming systems using locally-adapted species.

Introduce relatively small changes gradually. These will generally be more successful and more spontaneously adopted, than major radical changes introduced very abruptly (Bunch, 1995).

CONCEPT I

Old approach: **A topographic catchment ('watershed') is the most logical and optimum geographical unit for planning and implementing improvements in land use and management.**

New approach: **A community and the land it occupies and uses will often be the most practical unit initially for planning and implementing improvements in land use and management.**

BOX 10: A FARMER'S INFORMATION ON THE USES OF A RANGE OF PLANTS WITHIN A 1-HA COFFEE "SHAMBA", EKENYWA VILLAGE, ARUMERU DISTRICT, TANZANIA

SCIENTIFIC NAMES OF TREES AND SHRUBS	USES BY VILLAGERS														
	1	2	3	4	5	6	7	8	9	10	11	12	13	14	15
Acrocarpus flaxiifolia			+												
Albizzia lebbek		+													
Bombax sp.							+					+			
Bridelia micrantha			+						+						
Cassia siamea		+	+												
Cholorophora excelsa					+										
Commiphora sp.									+						
Cordia abyssinica			+		+			+							+
Croton microstachys								+	+						
Dracaena kilimanjarica						+		+							
Eucalyptus maidenii		+													
Ficus spp.		+	+						+						
Grevillea robusta		+	+		+				+	+					
Kigelia aethiopica				+										+	
Mangifera indica	+												+		
Melia azederach									+						
Morus alba	+								+						
Persea americana	+								+						
Rauvolfia inebriens		+						+			+				
Sorindea usumbarensis	+	+							+						
Syzigium cuminii	+														
Tamarindus indica	+	+							+						
	1	2	3	4	5	6	7	8	9	10	11	12	13	14	15

1 = Cattle boma; 2 = Brewing; 3 = Canoe; 4 = Fencing; 5 = Fodder; 6 = Fruit; 7 = Fuel; 8 = Furniture; 9 = Mattress; 10 = Medicine; 11 = Roof-beams; 12 = Shade; 13 = Timber; 14 = Tools, ploughs; 15 = Yokes

Comments

The topographic catchment may appear to be the logical area for planning and implementing improvements in land use and management from the viewpoint of those living downstream, especially in relation to the streamflow and accompanying sediments which are received downstream. However, it may not be the most logical or appropriate area from the viewpoint of those who inhabit the upstream areas, for the following reasons:

- The primary focus of development should be the well-being of the rural community rather than technical issues such as land use and soil conservation *per se*.

- The boundaries of rural communities are generally administrative and seldom coincide with topographical catchment or other landscape boundaries. Village areas define the extent over which a community can manage the land.

- In catchment planning emphasis is often placed on assigning land uses to land units in strict accordance with catchment topography and erosion hazard classification (i.e. according to land-use capability studies). This has generally resulted in proposals to re-organize the scatter of farms and fields into a supposedly more 'rational' pattern of land use. This has never been popular with farmers with small and often fragmented farms scattered across steep landscapes. Moreover, on steeplands all arable farming is often prohibited because it does not match the 'non-arable' classification of the land. Thus the implementation of land use capability studies is invariably difficult on steeplands. In contrast, changes or improvements in land management which do not require formal adherence to land capability classification, are generally more easily implemented, more effective, and much more acceptable to farmers.

- The primary purpose of catchment-specific improvements is related only to streamflow, water and sediment movements within and out of the catchment. If emphasis is shifted to better management of the soil in and around each farm in a village area, so that runoff and soil movement are largely eliminated, the primacy of the catchment basis diminishes.

- Improvement of soil quality and surface cover in any form of land use, will generally reduce soil degradation more effectively than relying solely on conservation-specific physical and vegetative structures laid out parallel to topographical contours.

- The adoption of any agreed changes in land-use or management practices will depend on the extent to which the changes are accepted by individual farmers, who will test them out according to their own criteria, initially on small plots, and then on a farm scale. Adoption may then spread to a community-area scale, and subsequently perhaps to a catchment-scale. It is therefore logical to commence implementing changes in land use and management at the farm level.

- For a given community, the appreciation of a physical catchment is often limited to the area within which other members' farms occur, and which can be readily seen within the valley. The location and extent of a community is more readily appreciated by farmers than that of a catchment.

Nevertheless, physical works aimed at controlling water flow - channels, ditches, banks, terraces, roads, and waterways - should always be laid out with reference to the topography of the catchment, rather than within individual boundaries alone.

Farm families and communities may become actively interested in joint action on a catchment basis when they realize that the volume and/or regularity of their water supplies are being threatened by land mismanagement, and/or that alterations in land use could improve the present situation. Nevertheless, they may not collaborate to this end until they have attended to their own individual and more immediate on-farm problems.

Principles for fieldwork

Work with the people of a community while thinking of catchments, rather than attempt to manage catchments for the benefit of people.

Do not insist from the outset that all land-use planning must be subordinated to organizing land use to fit the land-use capability classes (Plate 75). In offering advice and suggestions, think in terms of topographic catchments, and relate present and future water movements to catchment characteristics.

When mapping a village with the villagers, mark both the administrative boundaries which the villagers recognize, and superimpose the crests and drainage lines of the topographical catchment within which it lies, plus those of any mini- and micro-catchments that lie within the village boundary.

In situations where conservation-specific physical or vegetative measures are advisable, they should, wherever feasible, be laid out with reference to catchment topography, so as to limit the difficulties generally associated with the safe disposal of unavoidable runoff.

PLATE 75
A clearly-defined catchment, the upper part under Forest Reserve, the lower part evidently intensively used for agriculture, even though the land is very steep and would commonly be allocated (by technical specialists) to a mix of non-arable Land Use Capability Classes VI,VII and VIII. It is not feasible to suggest that the whole catchment should be closed to agricultural use. The pragmatic alternative is to help farmers to improve the productivity and water infiltration capacity of their land and, by so doing, to increase its stability and usefulness for the purposes they have chosen. Morogoro, Tanzania *[T.F. Shaxson]*

Chapter 6

Applications: planning, management and monitoring of steeplands

THE STEEPLANDS DILEMMA

In the majority of classifications of 'land use capability' or 'erosion hazard' for the tropical and subtropical areas, land over 12% slope - which has been termed 'steepland' in this document - is considered 'non-arable', and is recommended only for non-cultivation uses in permanent vegetation, such as native and/or planted grasses, shrubs and trees. Such recommendations are based on the implicit assumption that alternative equally-safe options for use are not usually available for cultivated steep slopes.

'Watershed planning' is usually based on the results of such classifications, in addition to those of land suitability for different uses, and commonly includes recommendations for maintaining or increasing the density of trees on such lands, starting at the upper end of the watershed and progressing towards the lower end.

However, increasingly large areas of 'non-arable' land continue to be brought into cultivation, particularly by resource-poor small-scale farmers, posing conceptual difficulties for making conservation-effective recommendations. For instance, if narrow-based conservation banks are to be suggested, extrapolation of the slope:horizontal spacing relationships, from flatter lands to steeplands, gives unacceptably-close spacing between banks and the loss of a fifth or more of the cropped area merely in the banks and associated channels. Since such works do not raise crop yields through any soil-conserving effects, farmers on steeplands are often unwilling to space them at the recommended short intervals. The result may be either that they are not installed at all; or they are spaced too far apart and cannot provide the degree of protection against runoff effects for which they were designed; or, if installed, they are not adequately maintained and fail in their supposed purpose. Bench terraces appear, from a technical viewpoint, to be the obvious answer to such difficulties. However, they are seldom implemented on steep slopes unless farmers feel there is a significant net benefit to be gained from the greater ease of management of high-value dryland crops or from the conservation of water for irrigated crops, which would compensate for the high expenditure of energy and/or cash used in their construction.

Standard texts on land-use planning and soil conservation may thus prove to be not particularly helpful to field staff faced with land degradation problems on steeplands. How can this problem be approached in a reasoned, effective and acceptable manner, while acknowledging that on steeplands the risks of degradation are higher, and that the stability of successful management is more fragile, than on flatter lands?

WHERE TO BEGIN

Three criteria serve as appropriate indicators of where to make a start:

• With those communities that have shown sincere interest in developing self-help initiatives concerned with land management and improvement, and which are requesting assistance with what they are trying to do. At such 'hot spots' there is likely to be widespread interest in the suggestions of field staff, a situation which facilitates working with the community (Shaxson 1989).

• In communities where the local infrastructure of roads, school and marketing is reasonable (or, at least, not so bad as to be the community's first and over-riding priority for attention). This allows early attention to be focussed on land-based problems which are near or at the top of the community's list of priorities.

• At locations which can be readily seen by passers-by, so that successful results of collaborative work can be easily seen and, hopefully, discussed by people from other communities.

HOW TO BEGIN

In terms of the use and management of the land within the boundaries of a particular community, day-to-day decisions are made by the members of the community; and thus their decisions affect the trends of change in the land's quality and productivity, stability and usefulness for chosen purposes. Therefore it is important for field staff to participate with the community (rather than the other way round) if solutions to problems are to be lasting in their acceptance and effectiveness.

'Village planning' is not merely producing reports and maps which express the villagers' 'wish-lists' of what they wish to do within the limits of their own understandings. The future stability of the village lands, and of the community that depends on the lands, will be closely linked with decisions made now about how future needs for water, fuel, food crops and marketable produce, space for building, etc. are to be produced from the community's strictly-defined land area, with its soils of whatever quality. These plans must take into account the increasing populations of people and animals, of demands for land and the diminution of farm sizes, and of declining opportunities for the adequate self-regeneration of soils and perennial vegetation.

In the spirit of participating with rural families in addressing problems related to land use and management, it is important for field staff (individually and collectively) to use participatory learning and action (PLA=PRA) methods from the outset, in order to:

• develop rapport and mutual trust;
• develop their own credibility (from nil at the time of first contact);
• listen and learn about people's livelihood systems within their community, their attitudes, preferences, cultural norms, cycles of activity, perceptions of their natural resources, etc.;
• become accepted as potentially useful sources of new technical information and ideas, and as participants in sorting out problems and possibilities;

- be able to function as catalysts in the introduction, assessment, adaptation and dissemination of new knowledge within the community.

Indigenous and local knowledge is very valid but may be limited in its ability to deal with rapidly changing situations. Outside knowledge can be important to rural communities in such situations (almost everywhere) by:

- reducing the time and money necessary for farmers to complete their own experimentation and assessments unaided;
- bringing in ideas and possibilities which have been developed outside the limits of their communication networks (Premkumar 1994:13).

SOME ASPECTS OF VILLAGE PLANNING

Appropriate land units

From the viewpoint of village farmers, attempting to deal with land-damaging problems by planning of the whole topographic catchment is an abstraction, and from the viewpoint of a thinly-spread advisory service, dealing with individual farmers is impracticable if widespread results are to be achieved. In most rural situations among small farmers, the most appropriate unit for collaborative work initially is the self-identifying community and the area of land to which the community has user rights.

The boundary of this 'social landscape' usually does not conform strictly, if at all, with the topographic limits of a catchment in which it is located. It conforms with the administrative/ political boundaries of its neighbouring communities.

Steeplands within the village boundary

The steepland portion of a village may support some of each of the whole range of agricultural land uses in the village, or may be dedicated to only a few specific uses, such as grazing, fuel production, etc. Improving the stability, productivity and usefulness of the steeplands must be considered not in isolation but in the context of the total village area. Therefore programmes of assistance to village communities should aim to attend to the community and their lands as a whole, rather than dealing only with their steeplands.

The greater the proportion of steepland within a village boundary, the greater the proportion of the overall area which is at risk of more rapid decline from poor management than would occur on flatter lands. This is a particular challenge in the face of the increasing needs to raise output per unit area in ways that will not increase land depletion or land degradation, and which will allow greater total output, greater diversification of land use, and hence enhanced risk-spreading, and more scope for adequate and assisted self-regeneration of damaged soils and vegetation.

The village as an ecosystem

Within the village boundary, the water supplies, land types, vegetation, livestock and people are interlinked and interacting such that the village as a whole is a dynamic ecosystem (Figure 13).

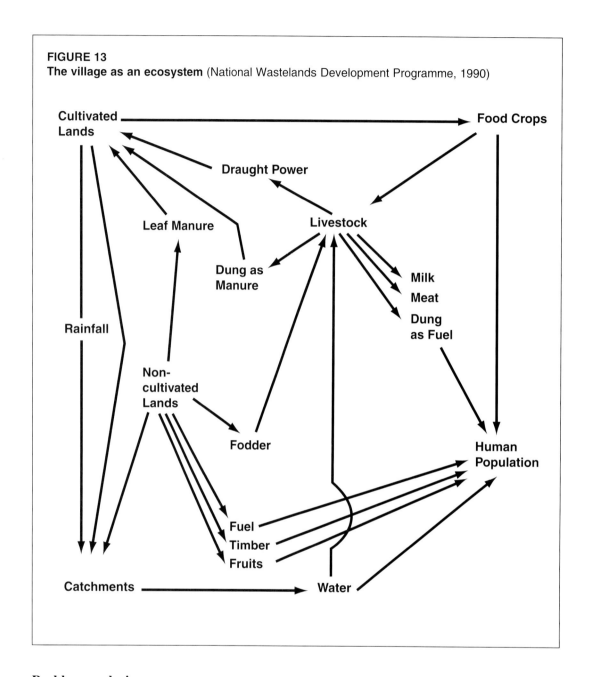

FIGURE 13
The village as an ecosystem (National Wastelands Development Programme, 1990)

Problem analysis

On steeplands in agricultural use, the detention-storage of potential runoff at the soil surface is less than on flatter lands, and the velocity of runoff is greater. Therefore the degree of replenishment of soil moisture reserves after dry periods is likely to be less on steeplands than on flatter lands, with increased risks of drought affecting plant yields.

Therefore more attention needs to be given to water capture and conservation on steep lands, rather than placing most emphasis on soil conservation. Increased rainwater infiltration is beneficial not only to annual crops but also to perennial vegetation. An example from the Siwalik Hills of north India shows the beneficial effects of increased moisture availability on the growth of young trees, with implications for increased production of leaves for fodder, branches for firewood, trunks for poles, etc. (Table 7).

TABLE 7
Increased soil moisture and growth of saplings (Mishra 1977, pers. comm.)
On upper slopes of Siwalik Hills, with generally well-drained soils, low 2 x 2-m earth 'boxes' were made around treated set of saplings, to retain rainwater. Increases in collar-girths and heights were compared with those of 'un-boxed' saplings over a 12-month period.

Initial plant material	+ boxes	- boxes
Tallplants (18 months old) Height increment Collar girth increment	2.08 m 11.1 cm	0.88 m 6.2 cm
Short plants (6 months old) Height increment Collar girth increment	2.03 m 11.4 cm	1.22 m 7.2 cm

In the same region, evidence from old people about the previous conditions in the landscape indicated that the present reduced quantities of water in springs and streams during the hot dry months of April-June are not so much the result of soil erosion (generally believed to be the major problem in the Siwalik Hills), but to declining rainwater infiltration. This led to an analysis of the contributing factors, which in turn pointed to the sorts of actions which could be undertaken by the village (Box 11 and Figure 14).

BOX 11: OLD PEOPLE'S MEMORIES AND LANDSCAPE CONDITIONS INDICATE A WATER PROBLEM

In an area of the Siwalik Hills in N India, old people reminisced about the dense forests and adequate year-long water supplies which they had known in their youth. A farmer pointed out where his father's land along the river bank had been 'eaten' by the river, whose width and propensity to flooding had markedly increased during the younger man´s lifetime. The size of boulders in the river bed and the evidence of flash floods were at odds with what old people remembered. These days, flow from hillside springs and seepages is greatly diminished, and even ceases altogether, during the three dry hot months before the monsoon arrives, in comparison with the situation earlier this century. Although there are no rain gauges or river-flow recorders and hence no long-term records in the area, these comments and observations clearly indicate a significant change in the surface hydrology of the area. This has evidently come about because of progressive changes in the partition of rainwater between soak-in and run-off, due to loss of cover on the surface and declining porosity in the soil's architecture. Erosion in the area is seen chiefly as degradation of drainage lines and streambeds, rather than as splash or rill erosion on the land surfaces.

Evidently, the answer to such problems lies in developing, with the inhabitants of the area, effective ways of increasing the infiltration capacity of the soils, rather than blaming 'soil erosion' and recommending implantation of soil conservation works as might otherwise have been the case.

In this situation, much of the apparent problem of reduced infiltration (estimated to be only about 60% of the mean annual rainfall of 1200 mm) has been caused by long-continued compaction and trampling by animals on the steep grazing areas, particularly those belonging to the government for which villagers have users rights, but which they do not fully control. Their feeling of lack of control leads to less care being taken in fire control of the forested grazing areas; low yields of trees, grasses and shrubs which result in animals spending long periods seeking food; relatively low output per head which forces farmers to own many animals to provide sufficient milk, dung or power; and large numbers of animals which lead to overstocking and greater soil and vegetation degradation.

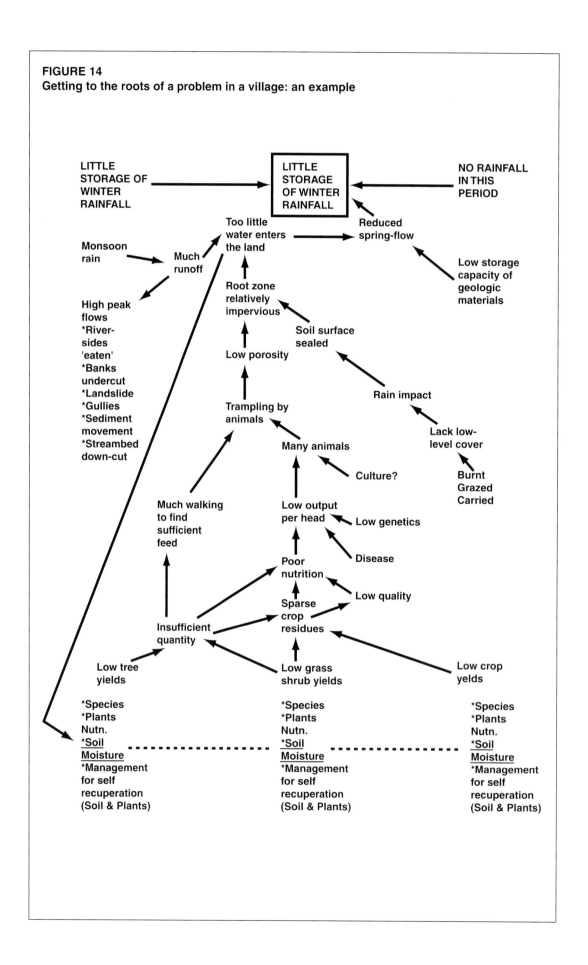

FIGURE 14
Getting to the roots of a problem in a village: an example

In collaborative village planning an important aspect of the work of field staff is to help identify and clarify such webs of interrelationships in the farming systems, and to suggest ways in which an overall problem (of which the above is an example) could best be tackled by inter-disciplinary attention to its interacting components.

In this particular instance, the possibilities for marketing milk led to the introduction of a cattle-improvement programme. Villagers with improved animals valued them more highly, and were not disposed to let them forage in a relatively uncontrolled manner across the steep country; rather, they have become interested in planting fodder for cut-and-carry feeding. As long as the number of animals does not increase, this leads to a lessening of the grazing pressure, with possibilities for soil regeneration and the improvement of superficial soil architecture and ground cover. Improvement of rainwater infiltration in such conditions is likely to diminish the length and severity of soil-moisture stresses, and so give higher production of grasses, trees and shrubs, thus reversing the downward spiral of declining productivity. Discussions were also beginning about the possible return of control of the grazing areas from government to the village community, which increased the community´s interest in taking responsibility for their care and management.

Mapping

Participatory learning and action methods encourage communities to make their own maps of many different aspects of their situation. Of particular importance to the improvement of land management are maps of both their own pragmatic classifications of land and the distribution of different land uses. Together with collaborative investigations in the field, such maps (which are not usually true to one scale across their surface) need to be compatibilized with more quantifiable images of the land such as air photos (variable but measurable scales according to altitude) and topographic line maps. Then the areas of different land classes (of suitability or hazard) may be calculated more precisely, giving information on the approximate amounts of each land unit within the village area. This information will then form the basis for planning.

Detailed instructions on how to undertake land capability and suitability classification in accordance with predetermined standard criteria are available in technical manuals in almost every country, often on the shelves of soil survey units or soil conservation departments. But these classifications, while being logically satisfying to technicians, may not accord closely with the field realities as perceived and experienced by farmers. Part of the work of field staff must be to combine the most appropriate parts of both systems into a usable framework that is valuable for the village situation.

Experience shows that farmers can often quickly interpret air-photo images in the light of their local knowledge, so that these can be used as a basis for semi-quantifiable mapping of farmers' information. The use of a simple soil auger during fieldwork with farmers may reveal the conditions of the soil below the surface, of which many farmers may not have been aware, but which may help to explain farmers' own observations about soil behaviour and soil productivity.

The resulting compatibilized maps – not only of the land's characteristics but also of its present uses and the way its tenure is distributed among different people and agencies – then become the basis for subsequent discussions among all concerned, about how to fit together the 'jigsaw puzzle' of future requirements, available resources, possibilities and restrictions which inhere in their particular village area.

Sequence of land planning activities

Without going into details of participatory village planning as a whole, suffice it to point out that a sequence of thinking provides the framework for discussions and actions (Vieira and Shaxson 1995):

- Participation with farm families and all other community members in defining, ranking and analysing their priority problems:

  ♦ seeking clues to the reasons for and causes of particular problems;
  ♦ collecting together, sorting, analysing, interpreting and clarifying relationships of data relevant to problems and potentials;
  ♦ characterizing the time dimensions of the problems: history; rates of change; sequences, etc.;
  ♦ identifying the important inter-disciplinary content and linkages of different aspects, in order to decide how to match the 'ecology of problems' with a relevant 'ecology of disciplines'.

- Selection of technical alternatives:

  ♦ identifying sources of information and assistance;
  ♦ choosing criteria for discriminating between alternative options;
  ♦ showing spatial and social dimensions of possible lines of action.

- Encouraging and assisting farmers' own processes of experimentation and adaptation to aid them in their decisions about what to adopt or reject:

  ♦ help with quantification of results;
  ♦ feedback of analysed data and interpretations to those who provided them;
  ♦ facilitation of farmer-to-farmer contacts and exchange of their own experiences.

In this whole process, field staff may act as 'agents provocateurs' in discussions, always mindful that any of their suggestions for discussion by the community should be conservation-effective, as well as being adequate, feasible, appropriate, economic, sustainable and acceptable. Questions may be formulated along the following lines:

- 'What are you going to do about ...?'
- 'If you do this, what about that...?'
- 'Why isn't the situation worse? What are the positive aspects which make the damage less severe than it could otherwise be?'
- 'What are the limitations to doing this ...?'
- 'What do you think about this ...?'
- 'What would be the repercussions of this action on people's daily lives (e.g. children's schooling; women's daily workload ...?'
- 'Where on your landscape could this best be done ...?'
- 'Who is going to do it, and who will maintain it ...?'
- 'How is it to be done ...?'

- 'What is the most appropriate sequence and rate at which the things you have decided will be done ...?'
- etc.

Non-conventional thinking – 'lateral thinking' - is valuable in such situations. The following checklist may be helpful:

- verify unsupported statements;
- seek alternative explanations of surveyed or received information;
- highlight possible interactions and synergies;
- pursue various implications of findings;
- seek out cause:effect relationships;
- examine, analyse and cross-check or verify one's own, and others', hidden assumptions and pre-conceptions;
- be ready for unexpected observations and answers by others.

HUSBANDRY AND STRUCTURES ON STEEP SLOPES

Achieving conservation by stealth

All aspects of good husbandry - especially the maintenance of adequate cover over and on the soil surface, and of good soil architectural porosity - are as applicable on steeplands as on flatlands. Where husbandry of crops, animals and land for production purposes is excellent, the conservation of water and soil is frequently achieved by stealth within the conservation-effective production system itself, without needing to rely upon add-on conservation-specific actions (Shaxson 1988).

Dangers

In steeplands the effects of gravity on downhill movements of water and soil over time are more rapid and severe, so that mistakes in husbandry are more quickly noticeable and more damaging in their effects than on lesser slopes. Both the implementation and continual maintenance of actions that are conservation-effective need more care and attention on steep slopes than elsewhere.

Cultivated fields on shallow soils in steeplands are more liable to rapid erosional damage because they will probably become saturated with water more quickly than deeper soils, and thus suffer surface runoff, rilling and gullying more frequently. Perennial vegetation of any sort on such areas will have the effects not only of binding the soil mass with their roots, but also of removing infiltrated water through evapotranspiration for longer periods of the year than can annual crops.

Contour actions

In all situations (except in the very few special situations where local experience has shown it to be undesirable) tillage on the contour is advantageous because the surface roughness plus the barrier effects of crop stems create temporary detention-storage of rainwater allowing more time to soak in. This practice complements the pre-requisites of good cover and excellent soil architecture.

Cross-slope lines of trees, grasses, stones, crop trash lines, banks and channels guide tillage along the contour, and can provide a basis for the passive formation of less sloping areas which may eventually (if desired) be formed into bench terraces.

Well-maintained lines of this sort may be favoured by farmers not so much for saving soil *per se* but for their effects in trapping applied inputs such as manure and seeds and preventing their loss down slope.

Because small farmers have multiple objectives but limited time, land and other resources, they prefer contour lines of those trees, shrubs and grasses which offer double advantages, i.e. (a) the production of timber, fodder, fruit, nuts, etc., and (b) land protection and improvement through the production of cover, litter or mulch and soil-architecture formation.

Harvesting water and soil

Even runoff water may be considered as an asset more than a hazard. Discussing with farmers the issues of water inputs, water storage in the soil, streamflow – and what affects the relationships between these factors – is often a key to promoting interest among farming families in planning how to catch and save rainwater where it falls.

Even water and soil on the move down slope on steep lands need not be wasted. Gravitational movement will occur anyway, even in geologically-slow terms. The challenge is to catch water and soil on the move, manage and make use of it, and when essential, let it pass on downhill before making use of it again e.g on a terrace immediately below (Box 12).

BOX 12: SOIL HARVESTING IN NEPAL

"The farmers are acutely aware of the loss of soil from their land, and seem to collect it continuously. ... In the monsoon, floodwaters deposit some soil on terraces, terrace slopes, in diversion and irrigation channels, etc. ...in the study area Districts near the river, farmers had made large silt traps to harvest soil. ... Visually the most impressive example of soil harvesting is in Bhujung. Ninety percent of the cultivated land comes from soil reclaimed from landslides and is built into terraces which occupy the lower fans of the landslides. The farmers explained that this method of collecting soil takes several generations".   (Tamang 1993:9-10)

Comparable actions are described from Karnataka in India, where farmers take a 'bottom-up' approach. They see erosion as an opportunity rather than a problem, and construct check dams to harvest soil and store moisture for crop-growing. Once the first check dam has filled with sufficient sediment for cropping, the next one is constructed a short distance upstream, and so on up the streamline, creating fertile micro-environments where none existed before (Premkumar 1994).

Physical works for steeplands

Considering land of any slope, there is a continuum of different types of physical cross-slope works from broad-based conservation banks ('broad-based bunds') on very shallow slopes to bench-type works on steeper lands (Figure 15).

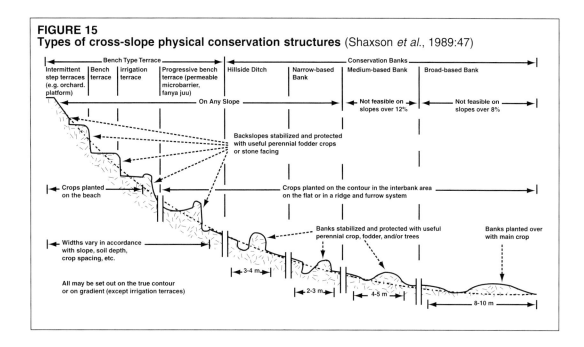

FIGURE 15
Types of cross-slope physical conservation structures (Shaxson *et al.*, 1989:47)

Broad-based and narrow-based banks, with channels on their uphill side, are advocated for slopes up to about 15%, but are much less suitable for steeper slopes. Here, types of bench terraces are more appropriate in that they provide more-or-less level platforms for crop growing. However, the work involved in their construction may greatly outweigh the possible benefits for many farmers.

The self-terracing effects produced by dense walls of grass strips or lines of stones – which rely on erosion processes to move the soil down slope to accumulate on their upper sides – can be augmented by manually throwing soil from below the line uphill, as in the 'fanya-juu' terraces of Kenya. Such terraces can be converted from outward-sloping (as they would be with only natural erosional processes) to inward-sloping terraces by moving soil from the inner side of the terrace to the outer edge, thereby altering the cross-terrace slope.

A backward-sloping bench terrace, of whatever size is markedly different in concept and function from conservation banks (bunds) with channels on the uphill side. The latter are intended to interrupt the downhill movement of runoff and soil arising in the cropped area above them, without the intention of making use of the runoff and eroded soil; the bench terrace aims to prevent undue movement of water and soil within the cropped area itself, and specifically allows use to be made of them. Thus bench terraces should be considered more as water-and crop-management devices than as soil-conserving devices.

Figure 15 shows variations in cross-sections of different types of cross-slope works; by contrast with conservation banks, bench-type structures may occupy all, some, or only a small part of a steep slope. If they are constructed for each individual large tree in a plantation – such as oil palms or citrus – they may form a series of individual platforms spaced intermittently both across and down the slope, and be called 'individual platform terraces'. They may be joined to form continuous terraces across the slope but be intermittently-spaced down the slope, as in 'intermittent orchard terraces', or the whole slope may be covered with terraces in both directions, as for tea, other closely-spaced crops, or irrigated rice (e.g illustrations in Shaxson *et al*. 1989:49-53; Hudson 1995:237-253).

Subsoil exposure

The formation of benches generally results in the exposure of lower-quality subsoils along the inner side of the terrace, nearest to the riser of the next terrace up slope, except in very deep, uniform soils such as those derived from volcanic ashes. To achieve an even crop stand across the terrace width, the exposed subsoil needs specific attention through the application of organic materials, tillage, fertilizer application and maybe differential cropping, perhaps using legumes for one or more seasons to regenerate a good rooting environment.

Disposal of runoff

In all cases on steeplands, the disposal of unavoidable runoff water (which even with good husbandry the soil has not been able to absorb) is a potentially dangerous and expensive problem, because the concentration of runoff into relatively narrow channels in natural but steep drainage lines produces very high mean velocities and great dangers of gullying. In steeplands this approach usually requires costly engineering-type torrent-control works (because the lowlands method of limiting velocity by making waterways wide and shallow is not practicable).

The most appropriate principle is to spread the runoff as widely as possible across the slope, as through the dense stems of contoured grass strips, or (as seen in areas of rice terraces), through the careful construction of a multitude of stone overfalls from one terrace to the next.

INDICATORS OF CHANGE

Farming-families' comments

In steeplands already in agricultural use, the hope is that they will remain, or become, stable as a result of good husbandry. It is desirable to know whether the actions implemented by village communities to improve the use and management of their lands are having positive effects, not only for their own satisfaction, but also so that decision-makers elsewhere – from farmers to politicians – will have evidence to guide their future decisions in other places.

There is much discussion in institutional circles about the needs for, and methods of, monitoring changes in natural resource conditions. Common to all suggestions, from interpretation of satellite imagery down to detailed laboratory analyses of specific soil components, are the enormous amounts of information required, for which the processing and interpreting can be very time-consuming, thus making it difficult to provide timely feedback and information to policy-makers and others guiding the development processes.

Monitored data, especially on changes in the conditions of natural resources and changes in people's activities, may have large margins of error ('background noise') related to uncontrollable variations in weather, politics, markets etc, and/or to the long time-scales over which the changes occur.

Qualitative information from a range of sources and using a range of indicators may prove to be just as reliable and useful as quantified information in the assessment of the nature, direction and rate of change of land qualities in an area (Casley and Kumar 1987).

Farming families use their own very pragmatic indicators of the conditions of their crops and soils in making day-to-day decisions about how they will use and manage their lands. Therefore, it makes sense to ask those who already live on the land, and who have continuous knowledge of what is happening, for information, including historical information, that can provide at least qualitative and perhaps semi-quantitative data, from which the more readily-perceived changes can be discerned. From these beginnings it can be decided what additional detail is needed to tighten up the information provided, and what complementary studies may also be required.

Farming families' comments may provide information on many different aspects of the changes they perceive (using their own indicators), plus their views on the explanations for the observed changes, for example:

- density of populations of people, and of animals, according to changes in land condition;
- patterns of out- and in-migration, and reasons;
- levels of satisfaction or dissatisfaction comparing present and past conditions of life in the farm or village community;
- changes in values of agricultural land;
- re-investment in agriculture or in non-farm activities;
- effects of drought periods on crop growth and stability of yields;
- productivity of crops, pastures, trees;
- incidence and severity of pests and diseases of crops and animals;
- diversification of on-farm enterprises;
- reasons for doing or changing farm operations in particular ways;
- availability, uses and values of favoured indigenous vegetation;
- springflow and streamflow characteristics;
- colours of plants and soils;
- output:input relations;
- soil physical conditions and the energy needed for tillage;
- additional costs of re-planting areas from which soil and inputs have eroded;
- spread of salinized areas;
- increases and decreases in runoff and soil erosion;
- disappearances, re-appearances of birds and other wildlife;
- frequency of burning residues, etc.;
- reasons for cutting/not cutting nearby forest trees;
- frequency of use of traditional practices for maintaining productivity, catching water, soil etc.;
- reasons for or against implementing currently-promoted recommendations;
- innovations taken up after hands-on training;
- farmer-to-farmer transfers and adaptations;
- farming systems, equipment types, etc.
- activity of worms, etc.

(For examples see also: Bhuktan *et al.* 1998; Wijewardene and Waidyanatha 1984:34; Hinchcliffe *et al.* 1995; Tamang 1993; Pretty 1994 (pers. comm.); Cassel and Lal 1992.)

At a field level it makes sense to start with characterizing the changes that farmers have begun to notice, then to consider related changes of which farmers may not be aware, and only then, when available resources have been adequately deployed in these ways, to start more speculative monitoring of less-obvious changes that may or may not be important.

If change in some aspect is not noted by farmers on the spot, it may not be very significant in the first instance, nor warrant much time spent on numerous and detailed analyses of a mass of data.

Farming-family members' statements on changes in output, food security, farming system etc. are comments about the impacts of various combinations of changes in social, economic and technical issues that may have taken place.

Within the area controlled by a community, the inhabitants' own comments are often important and illuminating concerning the nature and direction of change: such comments as:

"Things have/haven't changed; they are different/better/worse (a) than before, or (b) than I would have expected under the present conditions [of e.g weather pattern]."

In deciding which indicators may be the best to use, the spirit of co-participation is favoured (and suspicions of outsiders' motives are minimized) when the local inhabitants themselves are requested to suggest the indicators which they would use to characterize changes in the land that surrounds them, and in their own livelihoods. Such indicators are pragmatic, involve people's own self-interest, are likely to be enthusiastically monitored by those involved, and represent those things in the farmers' world which are changing most quickly or by the greatest amount.

Spontaneous farmer-to-farmer adoption

Farmers experiment with and maybe adapt new ideas to their individual circumstances before they decide whether or not to adopt them. An indicator of their enthusiasm for, and satisfaction with, some new method, which may include better land husbandry practices, is therefore the rate of spread through informal farmer-to-farmer contacts independent of any efforts by extension staff (see e.g. Bunch and Lopez 1998; Cheatle and Njoroge 1993).

FARMER SATISFACTION

In rural situations, farming-family members' attitudes, perceptions, observations, suggestions, knowledge and interpretations can be particularly important. It is their views of apparent benefits or disadvantages arising from changes in enterprises or technologies which will determine whether they will sustain the improvements in the future. Their comments may also indicate whether they are becoming more, or less, satisfied with their situation because of changes in the condition and quality of the land. Such comments may indicate a (human) aspect of sustainability, i.e. their expected permanence of residence in the area. If they are increasingly dissatisfied as time goes on, many will leave an area and the overall productivity will decline. If, as the land quality improves, they are increasingly satisfied, the land will be well husbanded and maintained.

Their degree of satisfaction is also a useful indicator of change in land quality, though once again the reasons for change and the nature of necessary improvements require more detailed analysis than a mere monitoring of the indicator.

Other people's observations

Other sources of relevant indicators of change can be sought from staff in both the technical and administrative sections of the district administration: advisory staff can provide information on commonly observed changes across the district as a whole, which individual farmers may not have noticed; the district roads engineer may comment on the increased or diminished costs of road maintenance due to changes in severity of runoff and erosion; the popularity of local political figures may be affected by their support for (or indifference towards) rural improvement programmes which have resulted in improved land quality; variations in the local water authority's costs of treating water for urban or domestic use may indicate changes in the sediment and chemical loads in the rivers from catchments upstream.

Photography

It is also important to add one's own observations and interpretations to what farmers say and show. In addition to note-taking, an ordinary hand-held camera is an important tool for recording observations which might otherwise be difficult to quantify or to describe adequately in words. Although it has great potential for recording serial changes at particular places, the simple camera is much under-used for this purpose. Photography merits greater use as part of the monitoring record. In addition to single photos in two dimensions, a pair of photos of the same scene, from viewpoints only a few centimetres apart, are in most cases sufficient to record three-dimensional views which can be recreated even with small and inexpensive viewers, either pocket stereoscopes or polarized spectacles.

Photographs can provide visual records of changes and differences over time, and from place to place, of physical features which otherwise would have no permanent record (see for instance Tiffen *et al.* 1994). Semi-quantitative comparisons may be possible from a carefully-planned series of photos of the same place at different times, and of a common feature in different places at almost the same time. Photo sequences over time may suggest what complementary indicators should or could be measured in future.

Scoring

Where quantitative data may be difficult to define and collect, systems of scoring of particular features provide an intermediate and valid step (e.g. Douglas 1995).

Quantitative measurements

For an overall initial assessment of changes in land quality, three broad types of key indicators may be important to complement farmers' own observations (Shaxson 1996):

* Above the soil surface, as related with yields:
    * **aerial cover**: its density, distribution, duration, timing;
    * **stress in plants**: growth rates; timing and frequency of wilting; visible nutrient deficiencies or imbalances.

- On the soil surface, as affecting particularly soil moisture, runoff and erosion:

    ◆ **contact cover**: its density, distribution, duration, timing;
    ◆ **porosity of at least the topsoil layers in millimetric bands**: proportions of incident rainfall becoming infiltrated (effective rainwater) vs. runoff (ineffective rainwater).

- Below the soil surface:

    ◆ **organic matter content and biological activity, as affecting multiple features**:
        - soil architecture:
            > structural stability;
            > gas exchange
            > water movement and retention/release;
        - cation exchange capacity:
            > nutrient capture and retention;
            > pH and nutrient buffering;
            > nutrient availability;
        - source of small amounts of recycled nutrients.

If greater levels of detail are required many soil attributes might be measured (Box 13).

BOX 13: EXAMPLES OF POSSIBLE INDICATORS OF SOIL QUALITY

(a) Soil attributes and soil quality (Elliott 1991)

Soil attributes	Elements of soil quality - ability to				
	Accept, hold and release nutrients	Accept, hold and release water	Promote growth	Provide a suitable soil biotic habitat	Resist degradation
Surface horizons					
organic matter					x
total	x	x	x	x	x
labile	x		x	x	x
nutrient supply	x		x	x	
surface soil texture	x	x	x	x	x
surface soil depth	x	x	x		x
surface soil structure		x	x		x
surface soil pH	x	x	x	x	
surface soil elect. conductivity	x		x	x	x
Limiting horizon					x
subsoil texture	x	x	x	x	x
subsoil depth	x	x	x		x
subsoil structure		x	x		x
subsoil pH	x		x		
subsoil elect. conductivity	x		x	x	x

(b) Physical and chemical properties at 3 depths
(Kiome and Stocking 1993)

Bulk density (g/cm³) :: Available water storage (v/v%) :: Sand(%) :: Silt (%) :: Clay (%) :: Texture (class) :: pH (H₂O) :: pH (KCl) :: Electrical conductivity (mmhos/cm) :: C (%) :: Cation exchange capacity (me/100g) :: Ca++ (me) :: K+ (me) :: Na+ (me) :: Base saturation (%).

Chapter 7

Implications

"Why is it that so many of those who should be achieving 'soil conservation' – farmers and other land users – seem unable to do so adequately? Is it because of: (i) lack of concern by farmers, policy-makers and the public at large?; (ii) inadequate government support – in terms of trained personnel, money and equipment to those who are 'trying to get the message across'?; (iii) lack of sufficient, or any, incentives for land-users to implement the already-available recommendations?; (iv) something wrong with the presentation of current recommendations?; (v) something wrong with the recommendations themselves? In the past, the first reasons have usually been used to explain any lack of success. This document suggests that we ought to be paying much more attention to the last two reasons in our urgent search for solutions."

(Shaxson 1987:11)

NEED FOR ALTERED CONCEPTS

In this document a number of concepts have been introduced which differ to some degree from what used to be 'conventional' thinking about land degradation, and in particular, that the main cause of land degradation was considered to be soil erosion.

Although this document is orientated to steeplands, the concepts presented are applicable to land of any slope. The urgency of their application is greatest on steeplands, not because steeplands are generically different from land of lesser slope, but because the effects of slope and gravity cause them to degrade more quickly than lands of gentler slope when all are subjected to a similar management regime that is not conservation-effective (Shaxson 1992a:24).

In the past, the answer to 'soil erosion' has been assumed to be 'soil conservation', which in reality had come to mean the addition of physical and agronomic conservation works into existing farm ecosystems, or even in some cases that existing land-use systems should be replaced by different ones in the name of 'wise land use'. It was assumed that a direct relationship of some sort exists between the severity of erosion and loss of productivity, the latter proportional to the former. But such a relationship has not been clearly demonstrated, despite much research. So the message 'soil conservation is good for you' has been based more on faith than on proof. Apparently logical means of implementing soil conservation have met with only sparse sustained success, despite huge investments of time, effort and money. Nevertheless the preoccupation with loss of productivity remains, and the conservation of soil productivity has become a key concern.

One of the key concepts is that land husbandry is what all concerned farmers do in order to look after their land - with greater or lesser success; it encompasses, but is not merely synonymous with, 'soil conservation'. What has been changing is not the belief that soil should be conserved, but the understanding of how that may best be achieved.

It is now becoming more widely accepted that sustainable soil productivity depends on the quality of soil as a rooting environment, and thus on the adequacy of its combined physical, biological and chemical qualities. The sustainability of its productivity depends in technical terms on its powers of auto-recuperation after damage, the potential for which resides primarily in its organic constituents and biological processes.

ALTERED EMPHASES

Changes in emphasis include both agro-ecologic and socio-economic aspects, each of which have implications for extension and advisory work, research, training, institutional arrangements, preparation of assistance programmes, legislation and policy-making. A few of the more significant shifts and their implications are indicated below:

Shifts in agro-ecologic emphases

- Husbandry of animals, crops and land interact one with another on a given farm in a system in which the farmer is the central figure in terms of management (Figure 16).

- In view of the serious effects of soil moisture stress on plant growth and yield, conservation of water as soil moisture is probably of greater concern and has more immediate impact than the conservation of soil particles.

- Rain**fall** in a vertical dimension precedes run**off** in a lateral dimension, and the mitigation of the damaging effects of intense rain**fall** deserves a higher priority than it has received to date in improving land husbandry.

- Accelerated erosion is a consequence of prior soil changes in cover and architecture, not a primary cause of land degradation *per se*; avoiding or remedying the causes of these changes should receive more attention than merely treating erosion as if it were the cause of the problem.

- Voids (spaces) which permeate the soil architecture and make the soil porous may, in terms of root growth and function, be more significant than the physical particles of soil; loss of soil voids is a prime factor leading to the initiation of runoff and erosion.

- A major consequence of the erosion of surface topsoil by splash and rilling is that underlying layers, generally less favourable for root growth and function, are exposed; loss of productivity will be minimized if swift attention is given to speeding up the restoration of the soil that remains behind, rather than imposing ever-denser networks of physical conservation works across the land surface.

- The practice of good land husbandry, together with that of crops and animals, achieves the conservation of soil *in situ* through the creation of good cover and porous soil conditions, complemented by the presence of any necessary more specialized conservation works designed to safeguard the land against infrequent but severe climatic events.

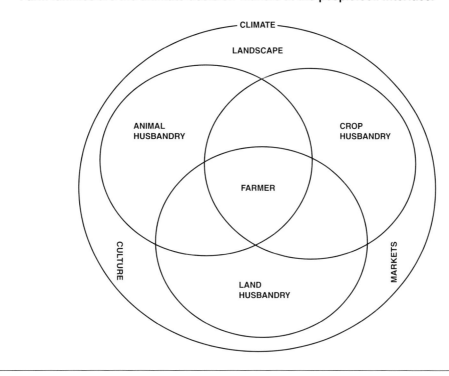

FIGURE 16
Three interrelated husbandries

- To maintain and improve productivity, stability and usefulness for the chosen purpose, land needs well-informed, active management over time.
- Farm families are the ultimate decision-makers at the people:soil interface.

Shifts in socio-economic emphases

- Farm families make rational decisions within the boundaries of their own limitations and potentials, and those negative and positive factors which they are powerless to modify: the challenge is to diminish the limitations to the benefit of their decision making, rather than attempting to change their rationality in the struggle for survival.

- The day-to-day life of farmers on the land and in the community provides them with their own observations, knowledge, perceptions and insights; on the one hand technical knowledge from outside the family may be able to enrich this information, while on the other hand they are able to enrich the understandings of non-farm agriculturalists who offer to assist them.

- Farmers make the ultimate decisions about how land will be used and managed; thus the opinions and decisions of farmers are the final filters through which variations in policies, recommendations, economic forces, legislation etc. have their ultimate effects on the land itself.

- Small farmers' decisions not to adopt particular recommendations, or to make unexpected responses to them, are generally based on careful assessments of the balances between risks, costs and benefits in the light of their objectives.

- Increasing awareness of these differences in concept from the conventional wisdom about 'controlling erosion' suggests that changes in attitudes may be needed more among non-farm agriculturists towards small farmers than among small farmers towards non-farm agriculturists.

When changes in these agro-ecologic and socio-economic factors occur together, a growing number of positive experiences are encountered - in both rural people's interest and enthusiasm, and in the improvements brought about in the land's condition (e.g. Moldenhauer and Hudson 1988; Hinchcliffe *et al.* 1995). These successes have implications for extension, research, training, institutional arrangements, legislation, assistance programmes, and policies, and for the interlinkages between them. Some of the implications are indicated below:

For advisory work

- Develop credibility in the eyes of farm-families so as to enable two-way communication of information and ideas, moving away from a top-down insistence that recommendations from higher authority should be obeyed.

- Accept the validity of farmers' experiences and observations based on their day-to-day familiarity with their surroundings.

- Assist farmers in identifying problems – hidden, looming in the future – in addition to those which are already obvious and perceived, and enable farmers to make their own decisions which are well-informed through effective advisory assistance.

- Help farmers address their 'ecology of problems' with assistance based on an adequate assessment of root causes, and an appropriate 'ecology of disciplines' provided by staff working together in an interdisciplinary manner to give integrated advice.

- Envisage the task of advisory worker more as catalyst + facilitator + travel agent + 'Yellow-Page' directory, and 'sprinkler of ideas' than as a transmitter of instructions from elsewhere.

- Assist farmers to achieve conservation by stealth through good husbandry and conservation-effective methods within their preferred farming systems, suggesting conservation-specific practices as and if necessary in the form of safeguards against severe but infrequent climatic events.

- In technical discussions, highlight the importance of soil moisture, soil porosity, soil architecture and optimizing the use of organic materials and the complementarity of mineral fertilizers as the foundations for a range of soil improvements and protection, including topsoil re-formation and auto-recuperation after damage or soil loss.

- Ensure a balance between women and men in advisory staff which is appropriate to the social and technical needs and potentials of the communities they serve.

- Treat advice on land husbandry and conservation of soil and water as a task involving a range of specialisms, rather than a job only for an isolated specialized department.

For research

- Promote investigations and research on micro-economic aspects of small farms to better understand the economic factors influencing farmers' decision making.

- Assess the probable direct and indirect costs and benefits deriving from research recommendations, to provide farmers with better information on which to base their judgements.

- By comparing local real-life on-farm problems with the availability of usable research recommendations, identify the gaps in farmer-usable information which require attention.

- Develop means of assisting farmers in their own research and experimentation, as part of an interdependent chain of research and adaptation activities from research plots to farmers' fields.

- Add 'conservation-effectiveness' as a formal criterion in 'sieving' research results into recommendations, so that this aspect becomes an integral part of a recommendation´s characteristics when the practice is properly applied by farmers.

- Investigate the occurrence, use and mechanisms of traditional land management practices which appear to be particularly constructed for, or effective in, holding water and soil in places where they are required, with a view to possible improvement in their effectiveness.

- Bring the management of organic materials and processes into the mainstream of soil-management research in view of the multiple benefits which they can produce in chemical, physical and synergistic aspects, with special reference to soil architecture, soil moisture and the efficiency of use of scarce fertilizer materials.

- Investigate the relationships between soil moisture and yield, with special reference to methods to increase the effectiveness of rainfall, that will probably complement the many past attempts at finding direct relationships between soil loss and yield.

- In view of the decline and disappearance of fertility-restoring long fallow periods, investigate ways in which soil quality for rooting, nutrient retention and water absorption can be maintained through the self-regeneration of topsoils that will counteract the detrimental effects of continual tillage and crop harvesting.

- Investigate reduced-tillage, minimum-tillage and no-till methods together with leguminous and other cover crops for their effects and potentials in conservation-effective production of preferred food- and cash-crops within common farming systems of different regions.

For staff training

- Extend training in participatory learning and action methods (successor techniques to PRA); stress issues and techniques such as the need for credibility, two-way communication in participatory and partnership modes, taking account of farmers' attitudes, local knowledge and expertise, rather than promoting only top-down and superior-technology approaches which appear to have hindered development in the past.

- Emphasize the inter-disciplinary linkages which exist and are to be employed in assisting farmers to achieve better husbandry of their land, crops and animals, and resist any suggestions that complex and interrelated problems of e.g. land degradation, should be tackled only on a mono-disciplinary basis.

- Introduce the ecological dynamics of land as a basis for arriving at concepts and actions for better land husbandry and the achievement of water and soil conservation as one of its consequences.

- Introduce training on aspects of participatory land assessment, classification, village and farm mapping, modifying conventional systems of soil survey and classification, planning of activities and watershed management so that they become more closely adjusted to the varied socio-economic realities of rural situations.

- Teach the elements of farm-level micro-economics, to engender better understanding of farmers' responses to new suggestions.

- Emphasize the four-dimensional nature of soil as a rooting environment, its heterogeneity in three spatial dimensions and its susceptibility to change over time - the fourth dimension, as a background to its appropriate care and management; include methods for the monitoring and feedback of information as a basis for continual land-quality assessment and the provision of optimal suggestions on land and soil management to farmers.

For institutions

- While retaining the essential technical disciplines and specializations in every institution, foster and ensure inter-disciplinary communication, collaborative arrangements and joint fieldwork within and between institutions so as to be more appropriate and responsive to the needs, problems and potentials of the small-scale farmer as seen from farm-families' viewpoints, rather than as imagined from the top downwards.

- Consider possible needs for either internal re-organization of existing institutions, or the re-combination of two or more institutions, to serve better the needs for appropriate assistance to farmers in multi-disciplinary aspects of better land husbandry, sustaining production and achieving conservation.

- Permit, encourage and facilitate adjustments of district-level arrangements within and between local offices of government and non-government institutions so as to address more effectively the potentials and limitations for better husbandry expressed by farm families and communities.

- Ensure that a competent advisory service on land husbandry issues is available to all land users, whether agricultural or not.

For aid programmes

- Move away from demonstration-type fixed-term field projects aimed at 'doing soil conservation', to long-term programmes of assistance to developing improved land husbandry and conservation-effective production in close collaboration and participation with rural communities.

- Work as closely as possible with or within government institutions in planning and running assistance programmes, so as to secure long-term continuity well after the cessation of outside support.

- Move away from ideas of pilot phases of rigidly-structured projects which aim to assess if a predetermined course of action is effective, towards phased programmes of assistance to better land husbandry with built-in flexibility to respond appropriately to farmers´ needs over time.

- Encourage long-term commitment on the part of the donor agencies to support the development process of change and adjustment **until** it becomes self-sustainable and self-supporting within the normal government structures. Although such commitments would be relatively-open-ended with regard to time, they could be controlled financially under a system of 'rolling budgeting'.

- With an understanding of the ecological dynamics of landscapes, it should be possible to foresee likely agro-ecological repercussions of various interventions, and build in suitable measures to avoid or contain them at the programme design stage, rather than wait until trouble has occurred and then attempt to undo the damage.

- All technical professionals whose work relates to land husbandry and conservation should assume responsibility for providing advice to planners and political/administrative decision-makers on the integration of agro-ecologic and socio-economic concerns into the design of policies, projects and programmes related to the improvement of rural areas and communities.

For legislation

- Local communities which accept responsibility from governments for the care and management of grazing lands, forests and other land and water resources, often wish to make local by-laws to provide controls, protections and directions; strengthen and support such initiatives which have local approval, by legitimizing them through national legislation.

- Where local communities have substantial shares in the benefits arising from outsiders using a local land resource - such as the entry of tourists into a wildlife management reserve - the community will become concerned for the areas' proper management in order to continue receiving benefits; in such situations promote legislation enabling the sharing of generated income between local people and the central treasury to benefit both management of the land and the wildlife. Beware of using coercive and punitive laws in relation to the use and management of land which may have the opposite effects to those intended, by antagonizing land-users and provoking destructive and spiteful reactions.

- Legislation which confirms long-term rights of access to, or ownership of, agricultural lands in rural areas can encourage farmers to invest time, energy and money into actions that will improve the health, productivity, stability and usefulness of the land. Such legislation will minimize the probability that the land will be mined of its fertility or that regular periods of soil recuperation will be neglected.

For policy-making

- Land degradation in rural areas has dire repercussions not only on those who live there but also on those who live downstream; maintaining the health, stability, productivity and usefulness of the land should be a primary issue in policies related to a nation's land, its use and husbandry; all other policies should support or contribute to that primary issue.

- Countries should produce and publish national statements about good land husbandry goals and how they can contribute to the conservation of the nation's land resources; these statements should serve as the basis for formulating policies on land use and management.

- All policies affecting farm families and their use and husbandry of their lands should be framed with due regard to the ways in which the policies may be perceived, and filtered, in land-users' decision-making processes, so that their decisions have desired favourable effects on the land.

- Economic policies should be framed with due regard to the fact that product-price instability plus inflexible input-cost structures lead to high-risk situations to which small resource-poor farmers may be particularly averse; if such situations prevail, farmers may be unwilling to respond to economic policy measures in the ways intended for the benefit of rural communities and the lands they manage.

- Policies on subsidies to farmers should ensure that farmers are recompensed fairly for any costs they are required to incur in producing off-farm benefits to other segments of society, and from which they themselves gain only partial or no return.

- Policies at national level should facilitate and support any appropriate changes in advisory work, research, training, institutions, programme planning, and legislation so that there is lateral compatibility and interdisciplinarity between institutions at the same level in a government hierarchy, as well as a compatibility between departmental policies, Ministerial policies, national policies and international agreements in the vertical direction.

- Policies of different agencies of government should be mutually consistent and also constant over time, to reduce some of the uncertainties which affect farmers' decisions on how to manage their land resources.

Conclusion

Re-thinking the concepts and approaches for helping farmers adopt improved land use and husbandry on steeplands has many implications - in extension, training, research, policy formulation, institutional organization, project formulation, legislation, and in the personal attitudes towards farmers of all those who work with, or for, the small-scale resource-poor farmer. Experiences with rural communities and farm families will indicate in more detail the needs and implications of particular situations. When successful local initiatives, supported by responsive field staff, take root in spite of the various hindrances posed by difficulties within existing institutions and services, it can be expected that resolving such difficulties will speed up the rate of adoption of good land husbandry throughout the region.

References

Anderson J.R. and Thampapillai J. 1990. Soil Conservation in Developing Countries: Project and Policy Intervention. (Exec. Summary of) *Policy and Research Series 8.* Washington D.C.: World Bank. ISBN 0-8213-1448-3.

Barber, R.G. 1998. Linking the production and use of dry-season fodder to improved soil conservation practices in El Salvador. *Proceedings of the 9th. International Soil Conservation Organisation Conference,* 26-30 August, 1996, Bonn, Germany. (In press).

Bhuktan, J.P., Basilio, C.S., Killough, S..A., De Los Reyes, M.F.L., Operio, S. and Locaba, R.V. 1998. Participatory upland agro-ecosystem management: an impact study. In: *Fertile Ground: Participatory Watershed Management.* London: Intermediate Technology Publications. ISBN 1-85339-3894.

Bonsu, M. and Obeng, H.B. 1979. Effects of cultural practices on soil erosion and maize production in the semi-deciduous rainforest and forest-savanna transitional zones of Ghana. In: *Soil Physical Properties and Crop Production in the Tropics.* Lal, R. and Greenland, D.J. (eds.). Chichester: John Wiley. pp. 509-519.

Bunch, R. 1995. *Two Ears of Corn: A Guide to People-centred Agricultural Improvement.* World Neighbours, Oklahoma, USA. 251 p.

Bunch, R. and López, V.G. 1998. Soil recuperation in Central America: measuring the impact four and forty years after intervention. In: *Fertile Ground: Participatory Watershed Management.* London: Intermediate Technology Publications. ISBN 1-85339-3894.

Buol, S.W. and Eswaran, H. 1993. Assessment and conquest of poor soils. In*: Proc. Workshop on Adaptation of Plants to Soil Stresses.* Univ. Nebraska, 1-4 Aug. 1993. INTSORMIL Pub. no. 94-2. Lincoln (USA): Univ. Nebraska. pp. 17-27.

Buol, S.W. and Sánchez, P.A. 1988. Soil characteristics and agronomic practices for sustainable dryland farming. In: *Challenges in Dryland Agriculture: A Global Perspective.* Unger, P.W., Jordan, W.R. Snead, T.V. and Jensen, R.W. (eds.). Proc. Internatl. Conf. Dryland Fmg, Bushland, Texas, USA. 15-19 August 1988. pp. 367-370.

Carson, B.R. 1985. Aerial photography as a base for watershed management at the village level in Nepal. In: *FAP Forestry Department Report on FAO/Finland Training Course in Forestry and Watershed Management for Asia and the Pacific.* Kathmandu. 19 May 1985.

Casley, D.J. and Kumar, K. 1987. *Project Monitoring and Evaluation in Agriculture.* Baltimore/ Washington D.C.: Johns Hopkins Univ. Press for World Bank. ISBN 0-8018-3616-6. 159 p.

Cassel, D.K. and Lal, R. 1992. Soil physical properties of the tropics: common beliefs and management restraints. In: *Myths and Science of Soils of the Tropics.* SSSA Spec. Pub. No. 29. Madison(USA): Soil Sci. Soc. Am. + Am. Soc. Agron. pp. 61-89.

Cheatle, R.J. and Njoroge, S.N.J. 1993. Smallholder adoption of some land husbandry practices in Kenya. In: *Working with Farmers for Better Land Husbandry*. Hudson, N.W. and Cheatle, R.J.). London: Intermediate Technology Pubns.+ WASWC. ISBN 1-85339-122-0. pp. 130-141.

Cintra, F.L.D., Mielniczuk, J. and Scopel, I. 1983. Caracterizaçao do impedimento em um latossolo roxo do Rio Grande do Sul. *Rev. Bras. de Ciencia do Solo*. 7.3, Sept/Dec. 1983. pp. 323-327.

De Bono, E. 1967. *The Use of Lateral Thinking*. Harmondsworth, UK: Penguin/Pelican Books. 141 p.

De Freitas, V.H. 1998. EPAGRI in Santa Catarina, Brazil : the micro-catchment approach. . In: *Fertile Ground: Participatory Watershed Management*. London: Intermediate Technology Publications. ISBN 1-85339-3894.

Doolette, J.B. and Magrath, W.B. (eds.) 1990. Watershed development in Asia: strategies and technologies. *World Bank Technical Paper No. 127*. Washington D.C.: World Bank. ISBN 0-8213-1606-0. 227 p.

Douglas, M.G. 1994a. *Environmental Management Guidelines for the Integrated Management of Mountain Ecosystems*. Draft for UN Environment Programme. Nairobi: UNEP. 60 p. [94/5/17].

Douglas, M.G. 1994b. *Sustainable Use of Agricultural Soils : A Review of the Prerequisites for Success or Failure*. Berne, Switzerland: Univ. Berne / Inst. Geog. 162 p.

Douglas, M.G. 1995. Guidelines for monitoring and evaluation of better land husbandry. *ENABLE - Newsletter of Assocn. for Better Land Husbandry (UK)*. No. 4, Nov. 1995.

Downes, R.G. 1982. *Institution-building for Soil and Water Conservation in Brasil*. Consultant's Report to Project FAO/UNDP/BRA/82/011. Rome: FAO/AGLS.

Elliott, C.R (ed.) 1991. *Evaluation for Sustainable Land Management in the Developing World*. Vol. 2. Technical Papers. IBSRAM Proc. no 12(2). Bangkok: Internatl. Bur. for Soil Res. & Mgt. 631 p.

El-Swaify, S.A. and Dangler, E.W. 1982. Rainfall erosion in the tropics. In: *Soil Erosion and Conservation in the Tropics*. Kussow, W., El-Swaify, S.A. and Mannering, J. (eds.). Madison, USA: Amer. Soc. Agron./Soil Sci. Soc Amer., ASA Special Publicn. No. 43. 149 p. ISBN 0-89118-068-0. [ex. Ch.1,2]

FAO. 1979. Yield response to water. J. Doorenbos and A.H. Kassam. *FAO Irrigation and Drainage Paper 33*. Rome: FAO. ISBN 92-5-100744-6. 193 p.

FAO. 1983. Guidelines: land evaluation for rainfed agriculture. *FAO Soils Bulletin 52*. Rome: FAO. ISBN 92-5-101455-8. 237 p.

FAO. 1991. World soil resources: an explanatory note of the FAO world soil resources map at 1:25 000 000 scale. *World Soil Resources Report 66*. Rome: FAO. 58 p.

FAO. 1995. Agricultural investment to promote improved capture and use of rainfall in dryland farming. *FAO Investment Centre Technical Paper 10*. Rome: FAO. ISBN 92-5-103630-6. 47 p.

Flaig, W., Nagar, B., Sochtig, H. and Tietjen, C. 1977. Organic materials and soil productivity. *FAO Soils Bulletin 35*. Rome: FAO. ISBN 92-5-100510-9. 119 p.

Fujisaka, S. and Garrity, D.P. 1991. Farmers and scientists: a joint effort. In: *Development of Conservation Farming on Hillslopes*. Moldenhauer, W.C., Hudson, N.W., Sheng, T.C. and Lee, S.W. Ankeny, USA: Soil and Water Cons. Soc. et al. ISBN 0-935734-24-4. pp. 221-232.

Gardner, R. and Jenkins, A. 1995. *Land Use, Soil Conservation and Water Resource Management in the Nepal Middle Hills*. Wallingford, UK: Inst. of Hydrology/London: Roy. Geogr. Soc. 8 p.

Garrity, D.P. 1991. Sustainable land use systems for the sloping uplands of southeast Asia. In: *Technologies for Sustainable Agriculture in the Tropics*. Madison, USA: Amer. Soc. Agron. (final draft). 43 p.

Gibberd, V. 1993. *Final Report: EMI Dryland Farming and Dryland Farming Applied Research Projects 1988-1993*. Chatham, UK: Natural Resources Inst./Nairobi: Ministry of Agriculture. 172 p.

Greenland, D.J. and Szabolcs, I. 1994. Preface to *Soil Resilience and Sustainable Land Use*. Proc. Budapest Symposium, 28 Sept.-2 Oct. 1992. Wallingford, UK: CAB International. ISBN 0-85198-871-7. 561 p.

Hall, G.F., Daniels, R.B. and Foss, J.E. 1982. Rate of soil formation and renewal in the USA. In: *Determination of Soil Loss Tolerances*. Amer. Soc. Agron. Spec. Pubn. No. 45. pp. 23-39.

Hallsworth, E.G. 1987. *Anatomy, Physiology and Psychology of Erosion*. Chichester, UK: John Wiley. ISBN 0-471-91212-3. 176 p.

Herweg, K. 1993. Problems of acceptance and adaption [sic] of soil conservation in Ethiopia. In: *Acceptance of Soil and Water Conservation: Strategies and Technologies*, Vol. 3. Topics in Resource Management in the Tropics. Baum, e., Wolff, P. and Zobisch, M. (eds.). Witzenhausen, Germany: DITSL. ISBN 3-9801686-4-6. pp. 391-411.

Higgins, G.M., Kassam, A.H. and Naiken, L. 1982. *Potential Population Supporting Capacities of Lands in the Developing World*. Rome: FAO + IIASA + UNFPA. ISBN 92-5-101304-7. 141 p.

Hinchcliffe, F., Guijt, I., Pretty, J.N. and Shah, P. 1995. *New Horizons: The Economic, Social and Environmental Impacts of Participatory Watershed Development*. London: Internatl. Inst. for Envt. and Devt. Gatekeeper Series SA 50. ISSN 1357-9258. 22pp.

Hudson, N.W. 1981. *Soil Conservation*. 2nd. edition. London: Batsford. ISBN 0-7134-3521-6. 324 p.

Hudson, N.W. 1987. Soil conservation in semi-arid areas. *FAO Soils Bulletin 57*. FAO, Rome. 172 p. ISBN 92-5-102606-8.

Hudson, N.W. 1988. Tilting at windmills or fighting real battles. In: *Conservation Farming on Steep Lands*. Moldenhauer, W.C. and Hudson, N.W. (eds.). Ankeny, USA: Soil & Water Cons. Soc. + WASWC. ISBN 0-935734-19-8. pp. 3-8.

Hudson, N.W. 1991. A study of the reasons for success and failure of soil conservation projects. *FAO Soils Bulletin 64*. FAO, Rome. ISBN 92-5-103087-1. 65 p.

Hudson, N.W. 1992. *Land Husbandry*. London: Batsford. ISBN 0-7134-5976-X. 192 p.

Hudson, N.W. 1995. *Soil Conservation*. 3rd. edition. London: Batsford. ISBN 0-7134-7353-3. 391 p.

IIED. 1994. *PLA Notes: Notes on Participatory Learning and Action*. Four-monthly Newsletter by Sustainable Ag. Programme. London: Intl. Inst. for Envt. and Devt.

Kiome, R.M. and Stocking, M.A. 1993. Soil and Water Conservation in Semi-Arid Kenya. Bulletin 61. Chatham (UK): Natural Resources Inst. ISBN 0-85954-359-5. 59 p.

Laing, D. and Ashby, J.A. 1993. Adoption of improved land management practices by resource-poor farmers. In: *Acceptance of Soil and Water Conservation: Strategies and Technologies*. Vol. 3 of Topics in Applied Resource Management in the Tropics. Baum, E., Wolff, P. and Zobisch, M.A. Witzenhausen, Germany: DITSL. ISBN 3-9801686-4-6. pp. 59-76.

Lal, R. 1981. Soil erosion problems on Alfisols in Western Nigeria : VI - Effects of erosion on experimental plots. *Geoderma* 25: 215-230

Lal, R. 1982. Effective conservation farming systems for the humid tropics. In*: Soil Erosion and Conservation in the Tropics*. Chapter 4. Madison, USA. Amer. Soc. Agron. Special Pubn. No. 43. SBN 0-89118-068-0. pp. 57-76.

Lal, R. and Pierce, F.J. 1991. The vanishing resource. In: *Soil Management for Sustainability*. Lal, R. and Pierce, F.J. (eds.). Ankeny, USA: Soil and Water Cons. Soc. + WASWC + SSSA. ISBN 0-935734-23-6. pp. 1-5.

Liao, M-C. and Wu, H-L. 1987. *Soil Conservation on Steep Lands in Taiwan*. Taipei: Chinese Soil and Water Conservation Society. 112 p.

Maene, L.M., Thong, K.C., Ong, T.S. and Mokhtaruddin, A.M. 1979. Surface wash under mature oilpalm. In: *Proc. Symposium Water in Agriculture in Malaysia*. Kuala Lumpur: Malaysian Soc. Soil Sci. 1979. pp. 203-216.

Moldenhauer, W.C. and Hudson, N.W. (eds.). 1988. *Conservation Farming on Steep Lands*. Ankeny, USA: Soil & Water Conservation Society. ISBN 0-935734-19-8. 296 p.

Moldenhauer, W.C., Hudson, N.W. Sheng, T.C. and San-Wei Lee. (eds.). 1991. *Development of Conservation Farming on Hillslopes*. Ankeny, USA: Soil and Water Cons. Soc. ISBN 0-935734-24-4. 332 p.

National Wastelands Development Programme. 1990. *Guidelines for Micro-Planning*. New Delhi.

Nicholson, M. 1971. *The Environmental Revolution*. London: Hodder and Stoughton. (Readers' Union edition.).366 p.

Norman, D. and Douglas, M. 1994. *Farming Systems Development and Soil Conservation*. Rome: FAO (Farm Management Series). ISBN 92-5-103448-6. 173 p.

Paningbatan, E.P., Ciesiolka, C.A., Coughlan, K.J. and Rose, C.W. 1995. Alley cropping for managing soil erosion of hilly lands in the Philippines. *Soil Technology* 8:193-204.

Pawluk, R.R., Sandor, J.A. and Tabor, J.A. 1992. The role of indigenous soil knowledge in agricultural development. *J. Soil and Water Cons.*, July/Aug.1992, pp. 298-302.

Pereira, H.C. 1973. *Land Use and Water Resources*. London: Cambridge Univ. Press. ISBN 0-521-09750-9.

Pereira, H.C. 1991. *Birth Rights*. SPA, Autumn 1991. pp. 21-23.

Piha, M.I. 1993. Optimising fertilizer use and practical rainfall capture in a semi-arid environment with variable rainfall. *Expl. Agric.* 29: 405-415.

Premkumar, P.D. 1994. *Farmers are Engineers.* Bangalore (India): PIDOW/MYRADA and Swiss Devt. Corpn. 40 p.

Pretty, J.N. 1994. *Soil Recuperation and Agricultural Regeneration in Central America.* Trip Report. London: International Institute for Environment and Development (IIED), Sustainable Agric. Programme. 3 p.

Pretty, J.N. et al. 1998. In: *Fertile Ground: Participatory Watershed Management.* London: Intermediate Technology Publications. ISBN 1-85339-3894.

Rist, S. 1992. Desarrollo y Participación : Experiencias con la Revalorización del Conocimiento Campesino, en Bolivia. *AGRUCO Serie Técnica. No. 27.* Cochabamba: Agroecología Univ. San Simón. 58 p.

Roose, E. 1993. Water and soil fertility management - a new approach to fight erosion and improve land productivity. In: *Acceptance of Soil and Water Conservation: Strategies and Technologies.* Vol. 3 of: Topics in Applied Resource Management in the Tropics. Baum, E., Wolff, P., and Zobisch, M.A. (eds.). Witzenhausen, Germany: DITSL. ISBN 3-9801686-4-6. pp. 129-164.

Roose, E., Dugue, P. and Rodríguez, L. 1992. La G.C.E.S. : Une nouvelle stratégie de lutte anti-erosive appliquée a l'aménagement de terroirs en zone soudano-sahelienne du Burkina Faso. *Revue Bois et Forêts des Tropiques.* no. 233, 3° trimestre, pp. 49-63.

Rosenberry, P., Knutson, R. and Harmon, L. 1980. Predicting effects of soil depletion from erosion. *J. Soil & Water Cons. (USA).* May/June 1980. pp. 131-134.

Schwab, G.O., Frevert, R.K., Edminster, T.W.. and Barnes, K.K. 1966. *Soil and Water Conservation Engineering.* 2nd. edition. New York: John Wiley. Lib. Cong. Cat. Card No. 66-14131. 683 p.

Scoones, I., Thompson, J. and Chambers, R. 1994. *Beyond Farmer First: Rural People's Knowledge, Agricultural Research and Extension Practice.* London: Intermediate Technology Pubns. ISBN 1-85339 250-2. 301 p.

Shaxson, T.F. 1971. Making better use of rain. In: *Proc. Sympos. on Water and the Tea Plant.* Kericho (Kenya): Tea Res. Inst. E Africa, 17-20 March 1971. pp. 177-193.

Shaxson, T.F. 1981a. Developing concepts of land husbandry for the Tropics. In*: Soil Conservation: Problems and Prospects.* Morgan, R.P.C. (ed.). Chichester, UK: John Wiley. ISBN 0-471-27882-3. pp. 351-362.

Shaxson, T.F. 1981b. Reconciling social and technical needs in conservation work on village farmlands. In: *Soil Conservation: Problems and Prospects.* Morgan, R.P.C. (ed.). Chichester, UK: John Wiley. ISBN 0-471-27882-3. pp. 385-397.

Shaxson, T.F. 1987. Changing approaches to soil conservation. In*: Soil Conservation and Productivity.* Proc. IV Internatl. Conf. on Soil Cons., Maracay, Venezuela, Nov. 3-9 1985. Pla Sentis, I. (ed.). Maracay: Soc. Venez. Ciencia del Suelo. ISBN 980-265-924-X (2 vols). vol. 1, pp. 11-27.

Shaxson. T.F. 1988. Conserving soil by stealth. In: *Conservation Farming on Steep Lands.* Moldenhauer, W.C. and Hudson, N.W. (eds.). Ankeny, USA: Soil & Water Cons. Soc. ISBN 0-935734-19-8. pp. 7-17.

Shaxson, T.F. 1989. *Achieving Sustainable Productive Land Use: A Framework for Interdisciplinary Field Action.* SPLASH (SADCC, Maseru) vol. 5, no.1 p.4. and no. 3/4 p. 4.

Shaxson, T.F. 1992a. National development policy and soil conservation programs. In: *Conservation Policies for Sustainable Hillslope Farming.* S Arsyad, S., Amien, I., Sheng, T. and Moldenhauer, W.C. (eds.). Ankeny, USA: Soil & Water Cons. Soc. ISBN 0-935734-28-7. 364 p. 14-25.

Shaxson, T.F. 1992b. Erosion, soil architecture and crop yields. *J Soil and Water Cons. (USA)* 47(6), Nov/Dec. 1992. : 433.

Shaxson, T.F. 1993a. Sustainability. *J Soil and Water Cons. (USA)* 48/4, July/Aug.1993.: 249.

Shaxson, T.F. 1993b. Conservation-effectiveness of farmers' actions: a criterion of good land husbandry. In: *Acceptance of Soil and Water Conservation - Strategies and Technologies.* Vol. 3 of Topics in Applied Resource Management in the Tropics. Baum, E., Wolff, P., and Zobisch, M. (eds.). Witzenhausen, Germany: DITSL. ISBN 3-9801686-4-6. pp. 103-128.

Shaxson, T.F. 1994a. Land husbandry's fifth dimension: enriching our understanding of farmers' motivations. *Proc. 8th. ISCO Conference.* New Delhi, 4-8 Dec. 1994.

Shaxson, T.F. 1994b. Introducción al concepto moderno de manejo integrado y conservacion de suelos. In*: Memoria del Taller sobre Planificación Participativa de Conservación de Suelos y Agua* in Paraguay and Brazil 21-30 June 1993. Proyecto Regional GCP/RLA/107/JPN: Documento de Campo 5. Santiago de Chile: FAO. p. 27-79.

Shaxson, T.F. 1996. Land quality indicators: ideas stimulated by work in Costa Rica, North India and Central Ecuador. In: Land Quality Indicators and their Use in Sustainable Agriculture and Rural Development. *FAO Land and Water Bulletin 5.* FAO, Rome, pp. 165-184.

Shaxson, T.F., Hunter, N.D., Jackson, T.R. and Alder, J.R. 1977. *A Land Husbandry Manual.* Zomba (Malawi): Govt. Printer. 627 p.

Shaxson, T.F., Hudson, N.W., Sanders, D.W., Roose, E. and Moldenhauer, W.C. 1989. *Land Husbandry: A Framework for Soil and Water Conservation.* Ankeny, USA: Soil & Water Cons. Soc. ISBN 0-935734-20-1. 64 p.

Siong, T.C. 1990. Indigenous conservation farming practices: Malaysia country review. In: *Indigenous Conservation Farming Practices.* ASOCON Report No. 7. Report of the Joint ASOCON/Commonwealth Workshop, Goroka, Papua/New Guinea, 3-7 Dec. 1990. London: Commonwealth Secretariat/ Jakarta: ASOCON.

Stallings, J.H. 1957. *Soil: Use and Improvement.* Englewood Cliffs, USA: Prentice-Hall. Lib. Cong. Cat. Card no. 57-05050. 403 p.

Stocking, M.A. 1996. Soil erosion: breaking new ground. In*: The Lie of the Land: Challenging Received Wisdom in the African Environment.* Leach, M. and Mearns, R. (eds.). London: International Africa Institute. Oxford: Currey, and Heinemann, New Hampshire, USA. ISBN 0-85255-409-5. pp. 140-154.

Stocking, M.A. 1995a. Soil erosion and land degradation. In*: Environmental Science for Environmental Management.* O'Riordan, T.O. (ed.). Chapter 12. Harlow: Longmans. ISBN 0-582-21889-6. pp. 223-242.

Stocking, M.A. 1995b. Soil erosion. In: *The Physical Geography of Africa*. Adams, W.M., Goudie, A.S. and Orme, A.R. (eds.). Chapter 8. Oxford: Oxford University Press. ISBN 0-19-828875-1. Pp. 326-341.

Stocking, M.A. and Sanders, D.W. 1992. The impact of erosion on soil productivity. In: *People Protecting their Land*. Proc. 7th ISCO Conference, Sydney, Australia, 27-30 Sept. 1992. Haskins, P.G. and Murphy, B.M. (eds.). Sydney: Dept. Cons. & Land Mgt./ ISCO. ISBN 0-7305-9977-9 (2 vols.) vol.1, pp. 102-108.

Tamang, D. 1993. Living in a fragile ecosystem: indigenous soil management in the hills of Nepal. *IIED Gatekeeper Series No. 41*. London: Intl. Inst. for Envt. and Devt. pp. 16.

Tato, K. 1992. The role of grassroots organisations in development - opportunity lost for soil conservation in Ethiopia. In: *People Protecting their Land*. Proc. 7th ISCO Conference, Sydney, Australia, 27-30 Sept. 1992. Sydney: Dept. Cons. & Land Mgt./ISCO. ISBN 0-7305-9977-9. vol. 2, pp. 435-443.

Tejwani, K.G. 1994. *Agroforestry in India*. New Delhi: Oxford and IBH Publishing. ISBN 81-204-0920-5. 233 p.

Tiffen, M., Mortimore, M. and Gichuki, F. 1994. *More People, Less Erosion: Environmental Recovery in Kenya*. Chichester, UK: John Wiley. ISBN 0-471-94143-3. 311 p.

Vieira, M.J. and Shaxson, T.F. 1995. *Criterios para la Identificación y Selección de Alternativas Técnicas para el Uso, Manejo, Recuperación y Conservación de Suelos y Agua*. Doc. de Campo 32 del Proyecto GCP/COS/012/NET. San José (Costa Rica):MAG/FAO; Rome: FAO (AGLS). 37 p.

Vogel, H. 1988. Deterioration of a mountain eco-system in the Third World due to emigration of rural labour. *Mtn. Res. & Devt.* 8(4): 321-329.

White, T.A. and Jickling, J.L. 1994. An economic and institutional analysis of soil conservation in Haiti. In: *Economic and Institutional Analyses of Soil Conservation Projects in Central America and the Caribbean*. Pagiola, L.E., Pagiola, S. and Reiche, C. (eds.). Washington D.C.: World Bank. pp. 98-106.

Whitehouse, I. 1990. Indigenous soil conservation in the South Pacific. In: *Indigenous Conservation Farming Practices*. Report of the Joint ASOCON/Commonwealth Workshop, Goroka, Papua New Guinea, 3-7 Dec. 1990. Jakarta: ASOCON; London: Commonwealth Secretariat /ADU. 53-66.

Wijewardene, R. and Waidyanatha, W. 1984. *Conservation Farming for Small Farmers in the Humid Tropics: Systems, Techniques & Tools*. Colombo/London/Eschborn: Min. Ag. Devt. & Research/Commonwealth Secretariat/GTZ. 39 p.

Willat, S.T. 1993. Soil and water conservation strategies in the South West Pacific. In: *Acceptance of Soil and Water Conservation: Strategies and Technologies*. Vol. 3 of Topics in Applied Resource Management in the Tropics. Baum, E., Wolff, P. and Zobisch, M.A. (eds.). Witzenahusen, Germany: Deutsches Inst. fur Trop. und Subtrop. Landwirtschaft/ DITSL. ISBN 3-9801686-4-6. 458 p. 193-212.

Williams, J.R. 1990. *The Erosion-Productivity Impact Calculator (EPIC) Model: A Case History*. Proc. Roy. Soc. Discn. Mtg. 7-8 March 1990. London: Royal Society. pp. 421-428.

Woruba, M.G. 1990. Some traditional soil conservation techniques in Papua New Guinea. In: Indigenous Conservation Farming Practices. Report of the Joint ASOCON/Commonwealth Workshop, Goroka, Papua New Guinea 3-7 Dec. 1990. Jakarta: ASOCON & London: Commonwealth Secretariat /ADU. pp 1-77; 41-44.

Annex 1

Slope units

Slopes are commonly stated in degrees or percentages, and the conversions between these two units are given:

Slope degrees/percentage conversions

Percentage	Degrees		Percentage	Degrees
8.8	5		5	2.5
17.6	10		10	5.42
26.8	15		15	8.32
36.4	20		20	11.19
46.6	25		25	14.02
57.7	30		30	16.42
70	35		35	19.18
83.9	40		40	21.48
100	45		45	24.14
119.2	50		50	26.34
173.2	60		60	30.58
274.8	70		70	35
567.1	80		80	38.4
vertical	90		90	42
	100		100	45

Annex 2

Delineation of slope classes

Topographic maps can be interpreted to provide information about the distribution and steepness of land of different slope classes, as described below. However, the validity and precision of the office interpretation must always be checked against slope measurements made in the field.

EXAMPLE OF HOW TO DERIVE AND MAP AREAS WITH SLOPES OF MORE THAN 12% STEEPNESS FROM A 1:15 000 TOPOGRAPHIC MAP

Calculation

a.   12% slope = 12 m vertical rise/fall over 100 m horizontal distance.

b.   For a vertical contour interval of 10 m on the map, a 12% slope will be represented by a horizontal distance between the 10 m interval contour of (10/12) x 100 m, equivalent to 83.3 m horizontal.

c.   At a scale 1:15 000, 10 mm = 150 m.

d.   Hence 83.3 m horizontal distance will be represented by (83.3/150) x 10 = 5.6 mm.

Procedure

1.   Draw a circle of 5.6 mm (say 6 mm) diameter on a small (5 x 5 cm) piece of flexible clear plastic (for other slopes, the circle would be of different diameter, calculated as above).

2.   Fix a sheet of tracing paper/film over the required area of the map, and run the circle between pairs of contour lines. Where the contour lines are closer than the circle's diameter, the slope is greater than the chosen slope limit (in this example: 12%).

3.   Colour in the areas where the contours are closer than the chosen slope limits.

4.   If several slope classes are to be marked on the same drawing, begin by marking the steepest slopes first, thus removing them from further consideration. If one were to begin with the gentlest slopes, a mass of overlapping colours would result.

Annex 3

How to visualize rainfall reliability in ten-day units (dekads)

1. Farmers have good knowledge of rainfall reliability at their farms, even if they do not express it in numbers.

2. Technical people tend to use numbers and diagrams to define parameters such as rainfall reliability, deriving knowledge from records of measured rainfall, rather than from first-hand experience.

3. It is not difficult to produce a reasonable pictorial representation of rainfall reliability at a place which has ten years or more of <u>daily</u> rainfall records:

   a. For the period covering the rainy season, divide each month into 'dekads', which are nominal ten-day units. e.g. January I, II, III; February I, II, III, etc. (Dekad III of each month may in fact have either 8, 9, 10 or 11 days according to the month).

   b. For each of the last ten years' rainfall records for the site, add up the total rainfall over the period of each dekad thus: Day 1+2+3+4+5+6+7+8+9+10 = Total, (e.g. for 1985 the ten-day total for dekad May I = 44 mm). Make a table as in the first example below, and fill it in as you go along. The Box shows such a Table fully completed.

   c. To see how little value is the mean for each dekad, add the ten-consecutive year total for the dekad and work out the mean. It can be seen from the example in the Box that in May I the mean was 6 mm, but the variation was from 0 to 44 mm. The mean value thus tells you little or nothing about what you might expect in the future at different levels of reliability.

   d. Now look at the second table, in which the figures have been interpreted as a basis for the diagram to be produced. For each dekad the number of occasions over the last ten years when increasing amounts of rainfall were actually exceeded is recorded (the rainfall-class differences could be 5 mm, 10 mm, or 15 mm intervals, - this example shows 10 mm intervals). Taking Dekad May I as an example, if you look at the column 1984-1993, you will see that 1 mm rain during that dekad was equalled or exceeded on 5 occasions (44, 1, 10, 1, 6 mm). However, 10 mm were equalled or exceeded on two occasions (44, 10 mm); 20 mm were exceeded on one occasion (44 mm), as were 30 mm (44 mm). However, none exceeded 50 mm. Compare dekad May I with say July II. In every year it rained something more than 0 in that dekad (72, 20, 47 ...17 mm). In nine of these years this dekad in fact had more than 10 mm (72, 20, 47, 30, 43, 64, 35, 140, 17 mm). In eight of these years rainfall equalled or exceeded 20 mm (72, 20, 47, 30, 43, 64, 35, 140 mm), and in seven of these years it equalled or exceeded 30 mm (72, 47, 30, 43, 64, 35, 140 mm). Five of those exceeded 40 mm (72, 47, 43, 64, 140 mm); and so on.

   e. Lines of equal frequency can be drawn directly onto the worksheet. Join all the '1's (interpolating between 0 and 2 in May II, and so on). This represents a frequency which you could expect 1 in every 10 years, or at 10% reliability for a stated amount to be equalled or exceeded (given in the left-hand column). Similarly, join all the '2's, to give a line for expectation at 20% reliability of rain of the stated amount; and so on for 30%, 40% ... 100%. The final result is seen in the last table.

   f. To find the period with say 90% reliability of 10 mm or more, look at the area <u>above</u> the '9' (90%) line, and read off the dekads in the year when 10 mm or more can be expected. In the example, this would encompass dekads June II, III, July I, II, III, and a later 'blip' at August II and III. To find 90% reliability of say 20 mm, the period would be restricted to July I and August II only. To find 20 mm or more with only 80% reliability, the period would extend from July I through to August III.

4. Although it is a somewhat crude method, it does allow you the understanding of a mass of highly-variable rainfall data. The pattern which emerges when the lines are drawn shows the peaks and troughs of rainfall during the season.

See below for the worked example.

EXAMPLE OF RAINFALL CHARACTERIZED BY DEKADS (MM RAIN PER DEKAD)

Month		May			June			July			August		
Dekad		I	II	III	I	II	III	I	II	III	I	II	III
(Yr)	19	0	0	1	2	24	29	24	72	13	50	40	25
	19	44	6	19	18	12	12	30	20	14	27	21	9
	19	0	0	49	1	102	39	32	47	81	22	45	26
	19	0	23	11	14	2	19	37	9	28	81	47	32
	19	1	3	59	0	46	26	111	30	134	5	29	14
	19	0	28	1	18	27	39	28	43	35	78	22	20
	19	10	15	40	2	25	32	71	64	75	5	52	72
	19	1	0	64	71	18	10	18	35	44	54	51	90
	19	0	5	36	13	11	21	32	140	50	77	39	97
	19	6	9	56	0	35	36	57	17	41	152	41	22
(Total)		(62)	(89)	(336)	(141)	(302)	(263)	(440)	(477)	(515)	(551)	(387)	(407)
(Mean)		(6)	(9)	(34)	(14)	(30)	(26)	(44)	(48)	(52)	(55)	39)	(41)

NUMBER OF YEARS IN WHICH A GIVEN AMOUNT OF RAIN WAS EQUALLED OR EXCEEDED IN THIS DEKAD

Rain mm equalled or exceeded	May			June			July			August		
	I	II	III	I	II	III	I	II	III	I	II	III
>1	5	7	10	8	10	10	10	10	10	10	10	10
>10	2	3	8	5	9	10	10	9	10	8	10	9
>20	1	2	6	1	6	7	9	8	8	8	10	8
>30	1	0	6	1	3	4	7	7	7	6	7	4
>40	1	0	5	1	2	0	3	5	6	6	6	3
>50	0	0	3	1	1	0	3	3	4	6	2	3
>60	0	0	1	1	1	0	2	3	3	4	0	3
>70	0	0	0	1	1	0	2	2	3	4	0	3
>80	0	0	0	0	1	0	1	1	2	2	0	2
etc.												etc.>

EXAMPLE OF RELIABILITY DIAGRAM FOR AMOUNTS OF RAINFALL EQUALLED OR EXCEEDED PER DEKAD

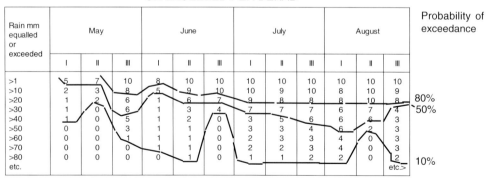

Annex 4

How to measure short-period rainfall intensity from an autographic rain chart

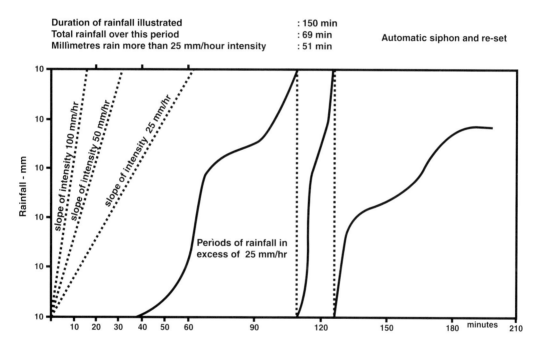

Duration of rainfall illustrated	: 150 min
Total rainfall over this period	: 69 min
Millimetres rain more than 25 mm/hour intensity	: 51 min

Automatic siphon and re-set

To assess the amount of rainfall of intensity greater than 25 mm/hr, place a sheet of tracing paper over the chart, mark the X and Y axes and the 25 mm/hr slopr. Then run the transparency across the chart and mark the beginning and end of each segment which has a slope steeper than that of 25 mm/hr. The add total mms which fell at high intensity.

Annex 5

Simple demonstration of the effects of cover and structure on runoff

Materials

1.   Five open-topped wooden boxes (or more if extra treatments wanted). WxLxH = 30x40x10 cm internally, one end 2 cm lower to provide a sill for runoff to flow over.

2.   Sufficient air-dry well-structured soil to fill all five (or more) boxes completely to 8 cm depth twice (enough to allow a second try if the first fails –see 9 below).

3.   One garden watering can (5-litre) plus coarse-sprinkler head, to simulate very heavy rain (make sure the water goes straight onto the boxes, not out sideways – a tin-can with one end off and the other with holes punched in it may give a better spray than the common wide-angle sprinkler end).

4.   Chopped dry grass or other mulch material sufficient to cover two boxes completely.

5.   A piece of wire gauze/mosquito netting 50 x 50 cm for use as a sieve for soil, and as a permeable cover over one of the boxes. (The mesh of the mosquito gauze should not be larger than about 2 mm, sufficiently fine to break up large water drops.)

Method

6.   Sieve soil until there is enough well-structured material, without roots, stones or fine materials, to fill one of the boxes 8 cm deep. Return the fines to the unused pile.

7.   Set aside enough well-structured soil to fill another box for a second try if needed.

8.   Pulverize the remaining material and sieve fine soil into the remaining three boxes, discarding any coarse materials.

9.   Prop up each box at an angle of about 30 degrees, each one a little apart from its neighbours, with the low sill of each box at the downhill end.

10. The four boxes with pulverized soil:

     ·   Leave one with no cover of any sort.
     ·   Fix the material used as a sieve over another box (without blocking the sill) to act as a means of breaking up large water drops before they reach the soil surface.
     ·   Cover the third box evenly and completely to a depth of about 5 mm with the chopped grass or other mulch material.
     ·   Cover the fourth box with a single layer of chopped grass, such that it is well-distributed but about 50% of the soil surface can be seen between the grass pieces.

11. The one box with well-structured soil is left uncovered.

12. Holding the watering can about 2 m above the first box, simulate a very heavy rainstorm by sprinkling one complete can of water evenly all over it.

13. Note the amount and colour of any runoff.

14. As soon as the runoff has ended, excavate the soil from the downhill half of the box by hand, noting the depth to which the water penetrated.

15. For each box, repeat steps 12-14, and discuss and take photographs of the results.

16. If the water has penetrated to the bottom of every box, and no differences can be discerned, empty out the wet soil and start again from the beginning, this time using say 3 litres of water instead of 5 for every box.

FAO TECHNICAL PAPERS

FAO SOILS BULLETINS

Availability: August 1999

Ar – Arabic   Multil – Multilingual
C   – Chinese   *   Out of print
E   – English   **   In preparation
F   – French
P   – Portuguese
S   – Spanish

The FAO Technical Papers are available through the authorized FAO Sales Agents or directly from Sales and Marketing Group, FAO, Viale delle Terme di Caracalla, 00100 Rome, Italy.